Disclaimer

The publisher of this book is by no way associated with the National Institute of Standards and Technology (NIST). The NIST did not publish this book. It was published by 50 page publications under the public domain license.

50 Page Publications.

Book Title: Best Practices for Modeling, Simulation and Analysis (MS&A) for Homeland Security Applications

Book Author: Sanjay Jain; Charles McLean

Book Abstract: This report recommends best practices for development and deployment of modeling, simulation and analysis (MS&A) tools for homeland security applications. The set of recommended practices applicable to any MS&A application includes: modeling practice, software engineering practice/ software reliability, model confidence/ verification, validation and accreditation, standards, interoperability, user friendliness and accessibility, performance, and innovative and unique elements. The recommended practices are discussed with respect to the MS&A application types for DHS that include: analysis and decision support, systems engineering and acquisition, planning and operations, and, training. The practices are also discussed with respect to MS&A application domains relevant to DHS.

Citation: NIST Interagency/Internal Report (NISTIR) - 7655

Keyword: Modeling, Simulation, Analysis, Homeland security, Best practices

NISTIR 7655

Best Practices for Modeling, Simulation and Analysis (MS&A) for Homeland Security Applications

Sanjay Jain
The George Washington University
Funger Hall, Suite 415
2201 G Street NW
Washington, DC 20052

Charles R. McLean
Systems Integration Division
Engineering Laboratory
National Institute of Standards and Technology
Gaithersburg, MD 20899

March 2011

U.S. DEPARTMENT OF COMMERCE
Gary Locke, Secretary
NATIONAL INSTITUTE OF STANDARDS AND TECHNOLOGY
Patrick D. Gallagher, Director

ACKNOWLEDGMENTS

The US Department of Homeland Security Science and Technology Directorate sponsored the production of this material under Interagency Agreement HSHQDC-08-X-00418 with the National Insitute of Standards and Technology (NIST). The development of this report gained from the time taken by a number of people at the organizations visited (see Appendix A) for providing the information of Modeling, Simulation, and Analysis (MS&A) tools developed and/or used by their respective organizations for homeland security applications. The support provided by Dr. Charles Hutchings, Deputy Director, Modeling and Simulation, Office of Test & Evaluation, Science & Technology Directorate of US Department of Homeland Security, through guidance on the content and in connecting with some of the organizations visited is also acknowledged. The work described was funded by the United States Government and is not subject to copyright.

DISCLAIMERS

The findings expressed or implied in this report do not necessarily reflect the official view or policy of the U.S. Department of Homeland Security, U.S. Department of Commerce or the United States Government.

Some software products may have been identified in context in this report. This does not imply a recommendation or endorsement of the software products by the authors or NIST, nor does it imply that such software products are necessarily the best available for the purpose.

Comments or questions about this report may be e-mailed to: simresponse@cme.nist.gov.

TABLE OF CONTENTS

EXECUTIVE SUMMARY 1
1 INTRODUCTION 7
   1.1   Study Methodology 8
   1.2   Organization of the Report 9
2 BACKGROUND 10
   2.1   Modeling, Simulation and Analysis 10
   2.2   MS&A Application Types 12
   2.3   MS&A Application Domains 15
   2.4   When should MS&A be used 17
   2.5   Current MS&A use 19
   2.6   Steps in a Simulation Study 20
   2.7   M&S Tool Development 21
3 BEST PRACTICES 24
   3.1   Conceptual modeling practice 24
   3.2   Innovative approaches 29
   3.3   Software engineering practices/ software reliability 34
   3.4   Model confidence/ Verification, validation and accreditation procedures 38
   3.5   Use of Standards 44
   3.6   Interoperability 47
   3.7   Execution Performance 51
   3.8   User friendliness and accessibility 54
4 RELEVANCE OF THE BEST PRACTICES TO MS&A APPLICATION TYPES 59
   4.1   Analysis and decision support 60
   4.2   Planning and operations 61
   4.3   Systems engineering and acquisition 62
   4.4   Training, exercises, and performance measurement 62
5 RELEVANCE OF THE BEST PRACTICES TO MS&A APPLICATION
DOMAINS 64
   5.1   Social behavior 64
   5.2   Physical phenomena 66
   5.3   Environment 66
   5.4   Economic and financial 67
   5.5   Organizational 67
   5.6   Critical infrastructure 68
   5.7   Other systems, equipment, and tools 68
6 APPLYING THE MS&A BEST PRACTICES 70
   6.1   TELL (Training, Exercise, and Lessons Learned) system 70
   6.2   Pandemic influenza impact study 71
   6.3   IMAAC/NARAC 73
   6.4   CIPDSS Decision Model 74
7 CONCLUSION 76
REFERENCES 77
Appendix A: Organizations visited for this study 87
Appendix B: List of Acronyms 88

This page left blank intentionally.

EXECUTIVE SUMMARY

The objective of this study is to define best practices for development and deployment of modeling, simulation and analysis (MS&A) tools for homeland security applications.[1] This study contributes to a higher level goal of encouraging improved practices for MS&A for homeland security applications. A number of efforts funded by US Department of Homeland Security (DHS) for the development and deployment of MS&A tools were surveyed via visits and interactions to understand the landscape of MS&A for homeland security applications. The survey was not intended to be exhaustive. It is hoped that this report will encourage a much larger effort to assess the multiple existing and under development MS&A application efforts funded by the U.S. government with respect to the best practices defined here and subsequently to improve the applications based on identified opportunities. It should be noted that for the purpose of this report, simulation refers to execution of computer models.

This study identifies several best practices. The set of best practices recommended for use for any MS&A application includes:
1. conceptual modeling practice,
2. innovative approaches,
3. software engineering practice,
4. model confidence/ verification, validation and accreditation (VV&A),
5. use of standards,
6. interoperability,
7. execution performance, and
8. user friendliness and accessibility.

Each of these practices is discussed using a common outline that includes: practice introduction, available guidance, recommended implementation, use for legacy vs. new applications, roles and responsibilities, costs/benefits, metrics and practice conclusion. Available guidance for the best practices ranges from a few research publications to many standards and policy documents. The recommended implementation identifies the guidance to follow for the cases where there are several competing options. The case for use of the best practices is supported by cost and benefits information where possible since limited documentation is available in literature on these aspects. The recommended implementation for each practice is briefly summarized below.

Conceptual modeling practice
Good conceptual modeling practice includes identification and clear communication of the intended use of the model. The intended use and the anticipated contexts should be clearly documented. Developers should utilize the principles of modeling provided by Pidd (1999), the methods of simplification provided by Zeigler (1976), and available modeling frameworks applicable to the problem under study. The selection of appropriate paradigm for modeling a problem should be determined only after careful consideration and not limited by the background of the modelers.

---

[1] Please see section 2.1 for definition of modeling and simulation (M&S) tools. For the purpose of this study, analysis tools are restricted to those that either employ M&S or are used in conjunction with M&S tools.

Innovative approaches
It is recommended that the developers and implementers of homeland security applications employing modeling and simulation focus on identifying innovative ways to meet the needs of the involved organizations and end users. The developers and implementers should identify the promising approach among the three suggested by Anthony et al. (2006):
1. The Back Scratcher: Make it easier and simpler for people to get an important job done.
2. The Extreme Makeover: Find a way to prosper at the low end of established markets by giving people good enough solutions at low prices.
3. The Bottleneck Buster: Expand a market by removing a barrier to consumption.

Software engineering practices/ software reliability
The MS&A tool development effort should go through a full software development cycle including planning, requirements analysis, design, coding, unit testing, and acceptance testing. In general it is recommended that the MS&A tools for homeland security applications be developed for usage across a number of jurisdictions. The development process for such an environment is generally recommended to be Capability Maturity Model Integration (CMMI) based. An iterative approach is recommended for development of the prototype versions of MS&A tools to ensure that the development meets the needs of the users.

Model confidence/ Verification, validation and accreditation procedures
The Department of Defense (DoD) verification, validation and accreditation (VV&A) recommended process guide (RPG) should be used for M&S applications for homeland security until guidance for such purpose is developed by DHS. It is recognized that while the verification and validation (V&V) steps in the RPG can be followed for homeland security applications, accreditation cannot be carried out until DHS identifies and authorizes agents for the purpose.

Use of Standards
Overlapping competing standards make it difficult for developers of MS&A applications to select the ones that the applications should comply with. The key idea is to ensure that the applications are compliant with relevant de jure and de facto standards. Ideally one would want the applications to comply with all applicable standards, but that would require prohibitive development costs. One should identify the standards being used by other applications that the new MS&A application will interoperate with to guide the identification of relevant standards. Error in choice of standards is a lower risk option than not choosing any and developing proprietary interfaces and operating practices.

Interoperability
The available standards and developing standards should be used up to the extent possible to improve the chances of integrating multiple homeland security MS&A applications that may be needed to study and analyze actual or potential events and responses. Development teams should strive for achieving higher levels of the conceptual interoperability model presented in Turnitsa (2005).

Execution Performance
Execution performance improvement of the simulation code execution is a complex task and should be implemented for applications that can lead to clear and significant benefits. Clearly,

the applications that are employed in a trans-incident operational setting can lead to significant benefits from faster execution and the ability to provide results faster. The efforts should follow the current de facto standards, Message Passing Interface for distributed memory platforms and OpenMP for shared memory processors, and thus reduce efforts required for performance improvement.

User friendliness and accessibility

The requirements for MS&A applications should consider the aspects of human computer interactions defined in ISO9241: Ergonomics of human system interaction (ISO 2001). Similarly, the development plans for the MS&A applications should comprehend the activities defined in ISO 13407: Human-centered design for interactive systems (ISO 1999) and ISO TR 18259: Ergonomics -- Ergonomics of human-system interaction -- Human-centered lifecycle process descriptions (ISO 2000). Special attention should be paid to the users for planning and operations and training, exercises, and performance measurement applications. The results should be presented to the incident management organizations using visualizations and terminology familiar to them and should include information on uncertainty associated with the results.

MS&A application for homeland security can be categorized using a grid formed using four application types and seven application domains as shown in Figure ES-1. The MS&A application types for homeland security include: analysis and decision support, systems engineering and acquisition, planning and operations, and, training, exercises and performance measurement. The seven application domains include: social behavior, physical phenomena, environment, economic and financial, organizational, critical infrastructure, and other systems, equipment, and tools. Each MS&A tool may address one or multiple cells in the grid. While relevance of the best practices could be discussed for each of the cells of the grid individually, for brevity the discussion is focused sequentially on each of the four application types followed by each of the seven application domains. Additional practices are identified by application type or application domain where applicable. For example, features providing realism of the visualizations make training applications of MS&A more effective.

Application Type ▶ / Application Domain ▼	Analysis and Decision Support	Systems Engineering and Acquisition	Planning and Operations	Training, Exercises and Performance Measurement
Social Behavior				
Physical Phenomena				
Environment				
Economic and Financial				
Organizational				
Critical Infrastructure				
Other Systems, Equipment, and Tools				

Figure ES-1: MS&A tools application types and application domains grid.

The relevance of the best practice to different application types is summarized in Table 1. Major aspects of the relevance assessments are discussed in Section 4 of this report.

Table 1: Relevance of Best Practices to MS&A Application Types

Application Type ▶ Practice ▼	Analysis and Decision Support	Planning and Operations	Systems Engineering and Acquisition	Training, Exercises and Performance Measurement
Conceptual modeling practice	Similar emphasis across application types			
Innovative approaches	Similar emphasis across application types			
Software engineering practice	Similar emphasis across application types			
Model confidence/ Verification, validation and accreditation (VV&A)	Focus on technical correctness for M&S; Different for analysis tools	Critical; Focus on technical correctness	Focus on technical correctness	Focus on realistic appearance
Use of standards	Emphasized for integration	Critical for operation use; Emphasized for planning	Emphasized for integration	Need to include compliance to Sharable Content Object Reference Model (SCORM)
Interoperability	Emphasized for inputs and outputs	Critical for operation use; Emphasized for planning	Emphasized for inputs and outputs	Needed but not as much as for decision support or operations
Execution performance	Emphasized for allowing exploring multiple options	Critical for operations use; emphasized for planning	Needed at a level to support process	Critical to present realistic time responses
User friendliness and accessibility.	Emphasized to support decision making	Targeted to incident management personnel	Low need due to highly skilled users	Emphasized to support trainees

The relevance of the best practice to different application domains is summarized in Table 2. Major aspects of the relevance assessments are discussed in Section 5 of this report.

Table 2: Relevance of Best Practices to MS&A Application Types

Practice ▼ / Application domain ▶	Social Behavior	Physical Phenomena	Environment	Economic and Financial	Organizational	Critical Infrastructure	Other Systems, Equipment, and Tools
Conceptual modeling practice	Important to select right paradigm	Important to select right paradigm	Important to select right paradigm	Need to emulate behavior and social processes	Need to emulate behavior and social processes	Important to select right paradigm	Important to select right paradigm
Innovative approaches	Similar emphasis across application types						
Software engineering practice	Similar emphasis across application types						
Model confidence/ Verification, validation and accreditation (VV&A)	Focus on Plausibility	Comparison with measurements	Comparison with measurements	Focus on Plausibility	Focus on Plausibility	Comparison with measurements; good data sources available	Comparison with measurements
Use of standards	Need Standards	Need VV&A standards	Helped by standards	Need Standards	Need Standards; guides available	Need standards for cross sector models	Need VV&A Standards
Interoperability	Need efforts	Need efforts	Relatively better; need efforts	Need efforts	Need efforts	Critical to integrate models	Needed to integrate in other models
Execution performance	Needed due to use of ABM	High Performance Computing (HPC) platforms used	High Performance Computing (HPC) platforms used	Needed due to use of Agent Based Modeling (ABM)	Needed due to use of Agent Based Modeling (ABM)	Needed for cross sector models	Not critical
User friendliness and accessibility	Needed to explain complex results	Helped by Geographical Information Systems (GIS) interfaces	Helped by Geographical Information Systems (GIS) interfaces	Needed to explain complex results	Needed to explain complex results	Needed to explain complex results	Not critical

This report does not single out individual MS&A applications as best practices. A MS&A application may use some or all of the best practices identified in this report. A few representative examples of MS&A tools and projects funded by DHS across different application types are discussed with the express intent to describe how the defined best practices may be used to help define the directions for further enhancements. The enhancements of the MS&A tools using the best practices defined in this report are anticipated to lead to their better use and to increased usage of MS&A tools in general across government agencies.

The report is intended for multiple audiences. Government program managers should find it useful in assessing the proposals and projects for MS&A for homeland security applications. Potential users within federal, state, and local agencies responsible for homeland security and associated functions should find it useful for evaluating and selecting MS&A tools. Developers of MS&A tools should find it useful in guiding their development efforts to follow best practices.

1 INTRODUCTION

This study is sponsored by the Office of Standards, Science & Technology Directorate of the Department of Homeland Security. The overarching goal is to improve homeland security via increased use of MS&A tools for improved analysis and decision support, systems engineering and acquisition, planning and operations, and, training, exercises, and performance measurement (please see section 2 for definitions of these application types). The objective of this study is to define best practices that are recommended for development and deployment of MS&A tools for homeland security applications and thus encourage increased and improved practices for employing these technologies across the department. It should be noted that the analysis tools included in the scope of this study do not refer to the whole gamut of such tools but only to those that either employ modeling and simulation (M&S) or are used in conjunction with M&S tools. The study should also encourage development of infrastructure required for MS&A including availability of standards, guidelines, and data sets. The study is not intended to be a survey of all relevant tools though a number of them were considered for understanding the current practices. It should also be noted that for the purpose of this report, simulation refers to execution of computer models and does not include role playing or live exercises by human subjects. The inclusion of training, exercise and performance measurement application type for this report is to address the majority of such applications that do use MS&A tools to support the live exercises.

The report is intended for multiple audiences. Government program managers should find it useful in assessing the proposals and projects for MS&A for homeland security applications. Potential users within federal, state, and local agencies responsible for homeland security and associated functions should find it useful for assessing such tools. Typically, the end users may be analysts supporting the agencies though in some cases the response personnel themselves may be the direct users. Developers of MS&A tools should find the report useful in guiding their development efforts to follow best practices. The developers may include personnel at government supported organizations such as the national laboratories, academia and commercial organizations.

The best practices defined in this document are from the perspective of MS&A tool development and deployment for homeland security relevant applications. Alternate ways of defining best practices could be from the perspective of promoting the use of MS&A (e.g., through providing guidance on policies and procedures, standard data sets, accreditation process, identification of gaps to guide development efforts, avoiding duplication of efforts, etc.) or from user perspective (e.g., input data analysis, data abstraction, ensuring right application of the model, user training, hardware & software availability such as for distributed parallel execution, etc). These other perspectives are related to the perspective taken in this report but are somewhat different.

The study developed the set of best practices based on literature surveys and discussions during visits to organizations involved in development and use of MS&A tools. The effort also built on several decades of combined experience of authors in MS&A development and applications in general including over a decade of combined experience in homeland security applications. It should be noted that current best practices may not be the same as ideal practices. An ideal application would be the one that exceeds the expectations on all the best practices specified. A current best application may be one that does well on multiple best practices among those specified. An application that does notably well on one criteria may also be highlighted as one to follow for that particular aspect. The set defined in this report aims to include ideal practices. It is understood that only a few projects may be able to follow the complete set of the applicable best practices and that the best practices may be "followed" to different levels.

The best practices should be followed in the context of a structured process for the development and deployment of M&S tools. Dr. Deming pointed out the need for guiding principles for best efforts as follows.
> "Best efforts are essential. Unfortunately, best efforts, people charging this way and that way without guidance of principles, can do a lot of damage. Think of the chaos that would come if everyone did his best, not knowing what to do." (Deming 1982, p. 19)

This report includes discussion of relevant processes for simulation study and M&S tool development to provide a context for best practices. Relevance of best practices is also discussed with respect to M&S application types and domains.

1.1 Study Methodology

The study team built on its knowledge of MS&A from its prior efforts in the area spanning several years since the first focused workshop on modeling and simulation for emergency response organized in 2003 (Jain and McLean 2003). This current study employed multiple ways of identifying MS&A tools that may be applied in the context of homeland security. These multiple ways included literature search, online search, participation in relevant workshops and conferences, discussions with DHS personnel, and visits to government and commercial organizations involved in use, development or support of MS&A tools. The intent of the visits was to review the tools and associated practices at leading organizations for developing and deploying MS&A tools for homeland security applications rather than conducting an exhaustive search.

The information collected from these multiple efforts was analyzed and used for defining the set of best practices in this report. The survey of academic literature is captured in Jain and McLean (2008). Appendix A lists the organizations visited for the purpose of the study.

This study approach was scoped to develop the set of best practices that are recommended for use. A much larger effort will be required if the study objectives were to assess all DHS funded ongoing efforts on multiple aspects, such as those addressed by the set of practices included in this report, and identify and document their practices. The assessments of individual efforts could then be followed by preparing a cross comparison to identify the high performers and best practices that may be transferable to other organizations. It is believed that the approach used for this report helps move significantly towards the same end goal, that of providing guidance for improvement of existing and new MS&A development and deployment efforts, at a much lower effort level.

1.2 Organization of the Report

The next section provides background information including definitions, when to use MS&A, current use of MS&A for homeland security applications, application types and application domains. It also presents steps in a simulation study and M&S tool development process to provide context for the best practices. The best practices recommended in this report are presented in section 3 in the rough order they may be employed during the development and application of MS&A tools. The MS&A tools and their use may vary widely based on the application type. Some of the best practices may be implemented differently based on the application type and application domain. Section 4 discusses the relevance of the best practices to MS&A application types, followed by section 5 that discusses the relevance of the best practices to MS&A application domains. Section 4 also mentions additional practices for MS&A application types where applicable. Similarly, section 5 presents additional practices for MS&A domains where applicable.

Following the description of best practices and their relevance to MS&A application types and application domains, the use of best practices is discussed with respect to a few notable MS&A tools for homeland security applications in section 6. The express intent of this section is to describe how the defined best practices may be used to help define the directions for further enhancements. The discussions suggest how the subject tool or project can employ best practices if it were not already doing so. No attempt is made to evaluate the individual tools or projects cited in this section since the defined scope did not allow for the effort that would be required for such an exercise. An attempt is made to select examples to cover the four application types. The last section concludes the report. As mentioned earlier, appendix A lists the organizations visited in the course of this study. Finally, appendix B lists the acronyms used in this report.

2 BACKGROUND

The discussion in this section provides background for presentation of best practices in the following sections. Many efforts identify use of M&S as a unit; however, it is a flawed process to develop models without regard for planned subsequent analysis (NRC 2006). This section defines modeling, simulation and analysis and briefly discusses why they should be used together for homeland security applications. It also includes the current understanding of how and when such tools are used and who are the current users for such a capability based on the interviews and data collected by the authors with support from the U.S. Department of Homeland Security.

2.1 Modeling, Simulation and Analysis

The terms modeling, simulation and analysis have been defined as below.

Modeling.
Application of a standard, rigorous, structured methodology to create and validate a physical, mathematical, or otherwise logical representation of a system, entity, phenomenon, or process. (DoD 1998)

Simulation.
A method for implementing a model over time. (Non-comprehensive examples: federation, distributed interactive simulation, combinations of simulations). (DoD 2009)

(1) A model that behaves or operates like a given system when provided a set of controlled inputs. (2). The process of developing or using a model as in (1). (IEEE 1990)

Modeling and Simulation (M&S).
The use of models, including emulators, prototypes, simulators, and stimulators, either statically or over time, to develop data as a basis for making managerial or technical decisions. The terms "modeling" and "simulation" are often used interchangeably. (DoD 1998)

The discipline that comprises the development and/or use of models and simulations. (DoD 2009)

M&S Tools
Software that implements a model or simulation or an adjunct tool, i.e., software and/or hardware that is either used to provide part of a simulation environment (e.g., to manage the execution of the environment) or to transform and manage data used by or produced by a model or simulation. Adjunct tools are differentiated from simulation software in that they do not provide a virtual or constructive representation as part of a simulation environment. (DoD 2007)

It should be noted that while the DoD definitions comprehend live exercises as simulation, this study is focused on MS&A tools that are software and computer based. The consideration of MS&A tools as being software is consistent with the DoD (2007) definition of M&S tools. Any MS&A tools used in support of training, exercises, and performance measurement will gain from the use of best practices discussed in this report. However, the practices discussed here do not apply to development of training and exercises that do not involve any MS&A tools.

Analysis has not been defined explicitly in relevant glossaries. Merriam Webster's Dictionary provides the following definitions of *analysis*: "Separation of a whole into its component parts," and "an examination of a complex, its elements, and their relations." In the context of this report, analysis can be defined as an examination of a complex system or process, its elements, and their relations to understand the potential outcomes and future states under a range of uncertain factors with the objective of assessing performance of the complex system or process or one of its major elements, or of comparing alternate configurations of the complex system or process or its major elements.

The term "analysis" is often used primarily in the context of decision analysis. Decision analysis has been defined as the discipline for helping decision makers choose wisely under conditions of uncertainty (Schuyler 2001). It involves structuring a decision problem into its major elements, the objectives, the decisions and alternatives, the uncertainties, and the consequences so that it can be analyzed (Clemen and Reilly 2001). Structuring the problem allows identification and implementation of appropriate decision models to assess outcomes of a decision and/or comparison of decision alternatives. The simulation output analysis is subsumed within the larger decision analysis context. Simulation is one of the alternate decision evaluation models whose outputs are analyzed as part of the decision analysis process.

It can be seen that analysis is a concept encompassing a wide variety of techniques for examination of a system, entity, phenomenon, or process. Modeling and simulation (M&S) comprises one set of techniques that can be utilized for analysis. A number of other techniques can be used for analysis including analytical models, diagramming techniques, statistical methodologies, experimentation, testing, data mining, etc.

For the purpose of this report, the term analysis refers to the techniques that can be used in conjunction with modeling and simulation. The analysis techniques may be used to develop insights into the potential outcomes and causes, e.g., for complex social behavior models, and/or they may be used for decision analysis.

M&S should be viewed as a methodology, that is, as a collection of related processes, methods and tools, rather than as a technique that is considered synonymous with method (INCOSE 2008). M&S encompasses a number of types such as discrete event simulation, continuous simulation, physics based models, etc. M&S can be applied for analysis and decision support as mentioned above, but it can be applied for other purposes such as training, exercises and performance measurements, and systems engineering and acquisition. For example, M&S can be used to present trainees with simulated dangerous situation that they may face during a terrorist incident and guide them through the recommended steps to follow. M&S may be used to create models of equipment that is being acquired to test proposed design alternatives. The boundaries of some of these application types are not defined in black and white. For example, some systems engineering applications of M&S may be seen as analysis. M&S use for different application types is discussed further in Section 2.2 and in more detail in Section 4.

There is value in bringing modeling, simulation, and analysis together for homeland security applications similar to what has been recommended for defense applications in a report from National Research Council (NRC 2006). The report states:

> "While many past studies examined defense M&S, this study examines military analysis as well, because all three activities—modeling, simulation, and analysis—are essential, intertwined inputs to decision making. Models (mathematical or otherwise logical representations of entities, relationships, or processes of importance to military operations) and simulations (exercises that include computations based on those models and possibly humans executing related tasks) are most effective when they are designed with analyses in mind. It is best for military analysts to provide input into the creation of the models and simulations that will underpin their analyses, and the best M&S-based support for decision makers comes about when M&S personnel and military analysts work coherently to explore and illuminate the issues facing those decision makers. Because of the need for such rich connections between M&S and military analysis, this report is written as though there were a single MS&A community, which would be a desirable situation.
>
> M&S is not, strictly speaking, a separate discipline that creates standalone tools, although there are certainly specialists within M&S who delve deeply to create new capabilities that can be incorporated into valuable tools. Similarly, military analysts are not always able to use canned software or predefined models if they are to perform well-targeted analysis."

Parallel to above recommendation from the National Research Council, homeland security analysts and M&S personnel need to work together for coherent application of these techniques to best support the decision makers with homeland security related responsibilities. Associated with the recommendation, the analysis tools discussed in this report will primarily include those that are integrated with or used coherently with M&S tools. Also, MS&A should be viewed as use of M&S for analysis in the context of the analysis and decision support application type.

2.2 MS&A Application Types

This report discusses relevance of the best practices with respect to MS&A application types in Section 4. An MS&A application is often developed to satisfy a single major objective. This characteristic identifies the primary purpose and thus the type of an MS&A application. Potential users looking for systems that are available to meet a specific need will typically want to search for an application based on its defined type or original intended use. The major categories of application types that have been identified are:

1. Analysis and decision support
2. Planning and operations
3. Systems engineering and acquisition

4. Training, exercises, and performance measurement

Each of the application types is briefly discussed below.

Analysis and decision support

Most applications of MS&A tools can be said to be for analysis and decision support in a general sense. However, for the purpose of categorization of the applications, this category is defined with a narrow meaning. This category refers to analysis and decision support that can be provided through use of such tools as: choice models, alternatives ranking models (e.g. analytic hierarchy process), information control techniques, analysis and reasoning techniques, representation aids, and human judgment amplifying and refining techniques. These tools may be seen as focused on analysis rather than modeling and simulation (M&S). They may be integrated with M&S tools to process the simulation results with various applicable analyses. These analyses help the decision makers gain insights provided by the simulation results and use them to make decisions. These tools may also provide a decision support environment utilizing M&S tools. The environment may provide interfaces suitable for a decision maker for executing simulation runs and guiding the analyses. M&S tools in this group may be used for reconstructing past incidents for analysis and improved understanding of the involved phenomenon. M&S tools that mimic human behavior may also fall in this category since they generally explore feasible outcomes and hence improve understanding rather than provide specific predictions.

Planning and operations

Operations applications are another rich area for employing MS&A tools. In particular, MS&A tools can be and are widely used for pre-incident operations planning, evaluating alternate strategies, policies and plans for response, recovery and mitigation. MS&A tools are also used to provide trans-incident operations support, though to a much lesser extent than operations planning.

The pre-incident operations planning area is perhaps the most appropriate area for utilizing MS&A tools with the present state of supporting infrastructure. Operations planning applications include evaluation of impact of natural hazards and man-made incidents, and the response, recovery and mitigation options to minimize the impact on the population and property from strategic to tactical levels. MS&A tools can be applied at a high level of abstraction and model long time periods in simulation to evaluate the strategic options, and they can be applied at detailed level modeling individual actions over a short time frame to evaluate tactical plans and procedures.

A large number of pre-incident operations planning applications of MS&A exist and perhaps a larger number are in various stages of development. The wide use of MS&A in this area is due to the current state of infrastructure required for their applications. MS&A tools typically require a large amount of data describing an incident and the location for modeling the impact and the response, recovery and mitigation options. At present, there is limited data that is readily available for MS&A. Use of an MS&A tool, thus, requires a substantial effort for collecting and cleaning up the data required. MS&A tools themselves need to go through verification and validation (V&V) process before they are used. The V&V process itself can be time consuming and may require a few iterations. The operations planning applications allow the time to go through data collection and V&V process for MS&A use. It is expected that as data sources and the MS&A tools mature, they will be increasingly used for trans-incident operational support applications.

The trans-incident operational support applications of MS&A include the tools that can be used to guide efforts during the response phase as the incident and/or its aftermath is unfolding. MS&A tools may be used for understanding the impact of an incident and to evaluate the response options. For example, MS&A tools may be used to estimate the areas that will be affected over time by a toxic plume emanating from an incident to guide the population evacuation efforts. They may be used to understand the impact of an incident on different infrastructure assets in the incident area and the cascading effect of the disruptions to guide the efforts to isolate and minimize such effects.

Systems engineering and acquisition

MS&A tools can be used to support the systems engineering and acquisition processes in organizations with homeland security related missions similar to their successful use for the purpose in DoD. Such applications include use of MS&A tools for: requirements definition, program management, design and engineering, efficient test planning, result prediction, supplement to actual test and evaluation, manufacturing, and logistics support. This application area primarily includes use of MS&A to evaluate systems and equipment through their design, development or manufacturing, and installation. The simulation models for evaluating performance of the system or equipment by itself are generally specific to the system or equipment being acquired and may not be data driven component models. These may include models of such aspects as mechanical strength and operation, chemical detection efficiencies, and electrical and electronic system operations. For example, the evaluation of a product design for detection of explosives in baggage would require modeling the physics of the process and would have to be specifically developed. On the other hand, models for evaluating the functioning of the system or equipment within the intended deployment environment may exist and be data driven. For example, the impact of the time taken for explosive detection system on the throughput of an airport security checkpoint may be done using generic airport passenger flow models configured using data to represent the facility of interest.

Training, exercises, and performance measurement

Perhaps the best known applications of MS&A tools are the applications for training, exercises and performance measurement. Such an impression may have been formed based on widely available reports of intensive use of MS&A tools for war games and associated training applications used by the military. MS&A tools allow creating realistic scenarios that a trainee may face in real life to test and improve his/her skills for executing his/her responsibilities. The responsibilities may be at the level of decision maker for directing preparedness, response, or recovery efforts or at the level of a first responder. The corresponding tools would vary from those that provide information from simulations regarding an unfolding incident to serious games offering first person interactions at the incident scene.

2.3 MS&A Application Domains

The report discusses relevance of the best practices with respect to MS&A application domains in Section 5. An application domain defines the type or types of behaviors, phenomena, processes, effects, etc. that are simulated within the application. Actual simulation implementations may model all or part of a single domain or multiple domains. For example, a traffic simulation might be considered a basic or homogenous simulation since it would focus on closely related elements, i.e., vehicles, roadways, traffic controls, and transportation support services. An evacuation simulation, on the other hand, might be considered a hybrid or heterogeneous simulation, as it might contain models of traffic, crowds, public transportation systems, and various organizations involved in the evacuation. In this document application domains will be described as basic simulations, rather than hybrids, as there are potentially infinite combinations of hybrid simulations that may be envisioned.

Major groupings of application domains for categorizing simulations include:
1. Social Behavior
2. Physical Phenomena
3. Environment
4. Economic and Financial
5. Organizational
6. Critical Infrastructure
7. Other Systems, Equipment, and Tools

Each of the application domains is briefly discussed below. Please refer to McLean et al. (2009) for more details.

Social Behavior domain

This domain includes individual and collective behaviors, movements, and social interactions between people at various locations of interest that are engaging in normal day-to-day activities or responding to an incident. Some examples of social behaviors that may be modeled include pedestrians in crowds, attendees at a public event, vehicle operators in traffic, carriers and transmitters of communicable diseases in public places, and consumers in stores. Some applications of this type of simulation include planning incident response operations (e.g., evacuations) or training incident management personnel.

Physical Phenomena domain

This domain encompasses the origin, propagation, and mitigation of various physical phenomena associated with emergency incidents. Examples of physical phenomena that may be modeled include earthquakes; explosions; fires; chemical, biological or radiological plumes; spread of airborne or waterborne disease and bio-agents; and biotic agents. These simulators may be used to support planning, training, as well as system engineering activities.

Environment domain

The internal and external environments that may be impacted by the occurrence of an emergency incident, propagate incident effects, and/or serve as the focus for response operations are included in this domain. Environments include the earth's atmosphere; watersheds and landmasses; ecosystems; indoor areas; and other confined spaces within and around man-made structures. These simulators may be used to support planning, training, as well as system engineering activities.

Economic and Financial domain

This domain includes the economic impact of an incident or policy at various levels including local, regional and national, and over various time horizons. Some examples of economic impact that may be modeled include exposure of the insurance industry to different disaster scenarios or estimating the effects of an incident on the local economy, and the time and resources required to recover to normal levels. Major applications of economic simulators include decision support, planning, and risk analysis.

Organizational domain

This domain includes the policies and procedures; activities and operations; decision processes, communications and control mechanisms; and information flows for various organizations and their members. Organizations of interest include those that perform incident management, support functions, or are impacted by incidents. Examples of organizations that may be modeled include fire departments, law enforcement agencies, health care institutions, government agencies, military units, businesses, voluntary assistance, and terrorist cells. Organizational models may be used for planning and risk analysis to evaluate effectiveness of organizations in dealing with various types of incidents. They also may be used to automate organizational operations in training simulations to minimize staff required for exercises.

Critical Infrastructure domain

The critical infrastructure systems, the impact of incidents on system elements, the propagation of incident effects on other interconnected, or nearby infrastructure elements, and the restoration of these systems after an incident are the subject of this domain. Examples of infrastructure systems include energy distribution systems, water supply, transportation networks, food supply chains, and communications networks. The most significant applications for this type of simulation are analysis of the risks associated with various types of disasters and planning of mitigation/recovery strategies.

Other Systems, Equipment, and Tools domain

This domain includes detailed operation and performance of various systems, equipment, and tools that are used in incident management, emergency response, and other homeland security related operations, or are affected by incidents and operations. The most significant application for this type of simulation is systems engineering. Examples of systems, equipment, and tools that may be modeled include: security scanners, sensors, and related systems; bomb disposal equipment; construction and firefighting equipment; hazardous material decontamination and disposal systems; search and rescue equipment; and various test equipment. Models may be used to support the development and enhancement of those systems, evaluate the effectiveness of those systems for specified purposes, and/or to provide detailed functional models for use in other simulations.

2.4 When should MS&A be used

MS&A can be used in four types of applications for homeland security discussed above. For each of these application types MS&A is one of a few to several techniques that can be applied for the purpose. Analysts should compare applicable techniques and select M&S only when it is the best suited for the purpose at hand. The applicable techniques may be compared on criteria such as feasibility, cost, and transparency.

A structured process is generally used for the analysis and decision support applications of MS&A, and may be applicable for some of the other application types too, particularly system engineering and acquisition. While a number of variations of decision analysis processes exist, they all have some common steps. Figure 2-1 shows a decision process adapted from Clemen and Reilly (2001). The third step of the process involves modeling the problem. This may include use of models in several ways including influence diagrams, decision trees, probability, hierarchical and network models, simulation models, and utility models. Admittedly some of these such as influence diagrams are not restricted to be "computer models" and can be at times solved manually. Also some of these models can serve to provide insights without being ever solved numerically.

Models may be used primarily for two purposes, evaluation and aggregations (Bouyssou et al 2005). The evaluation models are built to capture aspects of realities of the decision problem that are sometimes difficult to define with precision. The aggregation models are used to aggregate the results of complex evaluation models to help derive recommendations that take into account the preferences of the decision makers, and are robust in consideration of applicable imprecision, uncertainty and inaccurate determination.

M&S can be used in the role of evaluation models in the decision analysis process. That is, it can evaluate the performance of alternatives with respect to the objectives and thus help choose the best alternative. M&S should be used only when the analysts determine that it is the most appropriate technique to address the problem at hand. M&S is generally resorted to when the problem is too complex to be represented using mathematical programming models, includes many uncertainty factors that impact the outcomes of the decisions over a period of time, when the decision makers are involved in the process and are requesting a model that is transparent and easy to explain, and required resources are available.

M&S may also be used during the problem identification step though such use is generally not highlighted. Typically the same models that are used for problem identification are later enhanced and used as evaluation models later in the decision analysis process. The use of M&S as evaluation model then is more prominent at times overshadowing its use for problem identification. Use of M&S for problem identification is needed when it is not clear what the root of the apparent problem is. Clearly, M&S is only one of the techniques for the purpose and may be employed for complex problems. Generally techniques such as fishbone diagrams will be employed first to determine the root causes of an apparent problem.

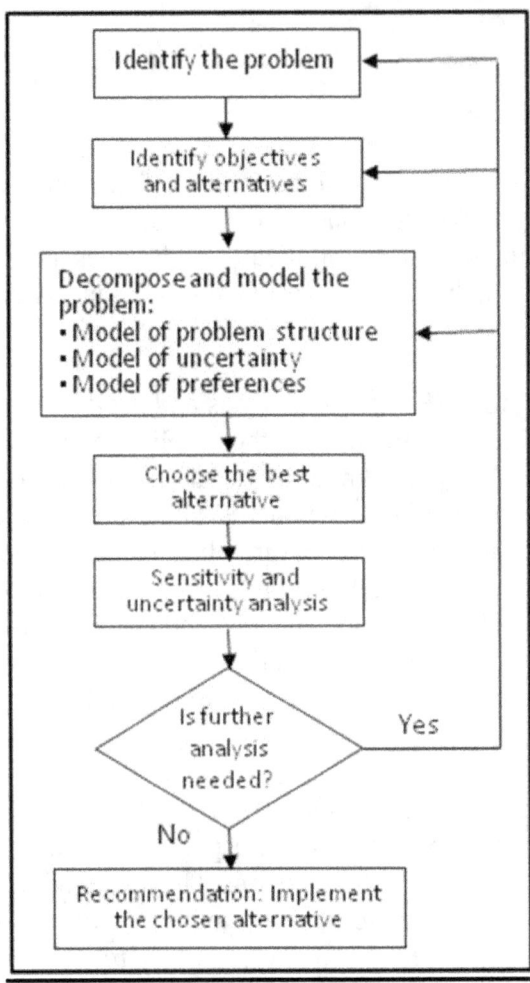

Figure 2-1: A decision analysis process flowchart (enhanced version of original adapted from Clemen and Reilly 2001)

Similar to the analysis and decision support application type, the other three application types have defined processes and alternate techniques available. The description of the processes and alternative techniques for the three applications types would digress from this document, hence brief examples are provided. M&S would be an alternative to experiments with physical prototypes for product evaluations within the systems engineering and acquisition application type. It will be an alternative to optimization based techniques such as mathematical programming and heuristics such as branch and bound, tabu search, etc for plan generations within the planning and operation application type. M&S is a cost effective alternative to running full scale live exercises with a large number of personnel and other resources within the training, exercises and performance measurement application type. It is also usually a more effective alternative for training outcomes analysis than techniques such as oral or written assessment of the trainees, and less expensive alternative than observation of trainees on the job or on test ranges and facilities.

2.5 Current MS&A use

The use of MS&A is one possible part of an overall problem solving methodology that may include identification of the overall problem, identification of appropriate approach, and implementation of the identified approach. The identification of the appropriate approach may lead to one or a combination of multiple approaches including experimentation, testing, analytical techniques, data mining, M&S, or M&S integrated with some analysis capabilities.

Modeling and simulation tools are being developed in an ad hoc manner in the U.S. Department of Homeland Security (DHS) to address specific problems identified by its Science and Technology Directorate based on inputs from "customers" such as first responders (Hutchings 2009). The ad hoc manner of development would consequently lead to ad hoc use of modeling and simulation tools. Limited efforts are focused on integrating analytical tools and data in a collaborative manner across the DHS enterprise. Some examples of ad hoc use of MS&A tools include use of atmospheric release dispersion simulations for prediction of plume behavior in the event of a toxic agent release, use of infrastructure simulations for predicting impact of hurricanes on coastal regions, and use of lexicographic analysis technique for allocation of budget to risk mitigation programs (Glickman 2008). A majority of the applications of MS&A tools appear to be in training area, in particular, associated with large exercises. For example, TOPOFF 3 (Top Officials) exercise utilized plume simulations to help guide the response efforts following simulated releases. Some applications of MS&A tools for actual incidents also exist particularly for the plume simulation and infrastructure simulation capabilities mentioned above.

Application of MS&A tools may require various personnel. A number of roles can be defined for MS&A for decision support. The NRC Report (NRC 2006) mentioned above identifies the following roles to support use of MS&A for decision support:
- *Analysts* create the formal model representation from the real-world problem, act as domain experts, and interpret and present results.
- *Modelers and programmers* translate the representation into a documented and executable form.
- *Implementers* develop and execute the experimental plan and transform the model's raw outputs into useable results.
- *Managers* oversee the MS&A team by managing personnel, making required purchases necessary for efficient operation of the team, checking for quality of the MS&A product, and interacting with others involved in M&S governance.

- *Consumers* employ MS&A to support decisions.

Multiple roles may be assigned to same people for studies with small scope. On the other hand, same role may be shared by a large number of personnel for studies with large scope. The role of modelers and programmers can vary widely based on the study scope. They may be able to use existing off-the-shelf software to model the problem or use a wide range of tools to build the models including spreadsheet software, computer aided design software coupled with physical or numerical models, or discrete event simulation models. They may require special technical expertise such as aircraft survivability following different damage scenarios.

The current users of MS&A capabilities are varied based on the ad hoc manner in which they are developed and deployed. A majority of the development efforts appear to be carried out by national laboratories. Such efforts are also being carried out at commercial contractors and at DHS funded university centers. All the roles defined above except consumers are usually residing at the developing organizations. The role of manager may be split between the managers at the developing organization and the program or project manager at DHS. The DHS program or project manager (PM) facilitates the identification of the problem, identifies the potential development organizations through an open announcement and oversees the selection of one for the purpose. The DHS PM then oversees the execution of the contract and may serve him/herself or bring in experts to check for the quality of the M&S product to the extent possible. The consumers of the MS&A outputs include responding agencies at federal, state, and local levels. For example, the outputs from simulations on impact of hurricanes on infrastructure may be used by Federal Emergency Management Agency (FEMA) and affected state and local jurisdictions to mobilize resources for anticipated response operations. Similarly, the outputs from plume simulations may be used by local response agencies to plan evacuation efforts.

2.6 Steps in a Simulation Study

Once the decision is made to use MS&A for a particular application, simulation models should be developed using a structured approach. A number of similar approaches for conducting a simulation study are available in the literature. Figure 2-2 presents a set of steps defined by Banks et al. (2009). It should be noted that while the steps were defined for discrete event simulation models, they would also apply to other kind of models such as continuous simulation, agent based simulation, etc.

The steps are grouped in three stages, model development, simulation runs, and output implementation. The model development starts with problem formulation and is followed by setting of objectives and overall project plan for conducting the simulation study. As mentioned earlier in the discussion of the decision analysis process, M&S may be used for problem formulation itself, in which case a mini-cycle of model development and simulation runs may have occurred beforehand. The problem formulation and objectives drive the development of a conceptual model and collection of data required for the model. The conceptual model and the data are used to translate the model into a computer recognizable format, i.e., it is coded using simulation software or a general purpose language. The model is then verified and validated before proceeding to its use.[2]

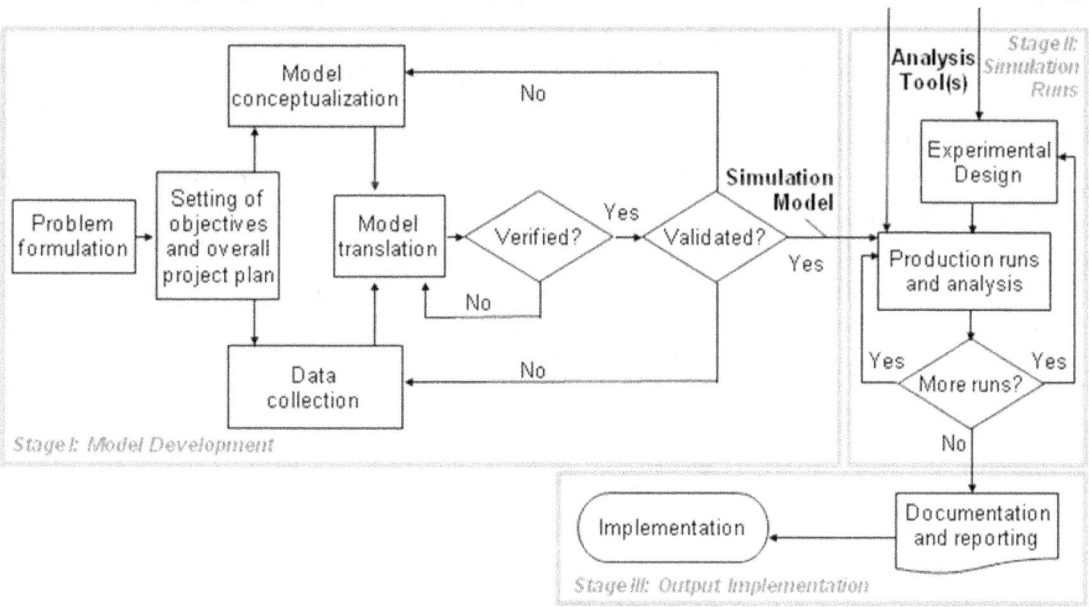

Figure 2-2: Steps in a Simulation Study (enhanced version of original adapted from Banks et al. 2009)

The model development is followed by simulation runs where an experimental design is used to determine the number of configurations and runs to be executed. The runs are then executed and the output analyzed. The analysis may identify the need for further runs. Once the requisite number of runs has been made and the results analyzed, the team can move to output implementation. The model and results are documented and reported to the consumers who may then proceed with implementing decisions based on the results of the analysis of simulation outputs.

2.7 M&S Tool Development

The steps shown in figure 2-2 have been proposed for a simulation study for addressing a specific problem. If the model is data-driven, i.e., coded to allow its configuration based on data,[3] it can be reused for addressing the identified problem in different settings. That is, the model is developed as a software that can be used for simulating a defined set of problems for multiple instances described by the data. For example, the plume models can be used for simulating release of various agents at different places under varying weather conditions by

---

[2] *Validation.* The process of determining the degree to which a model or simulation and its associated data are an accurate representation of the real world from the perspective of the intended uses of the model.
Verification. The process of determining that a model or simulation implementation and its associated data accurately represent the developer's conceptual description and specifications (DoD 2009)
[3] At times, modelers differentiate between parameters and data. Parameters generally refer to the inputs to the model that either largely determine the configuration of the model or their values have large influence on the performance of the modeled system.

describing all these aspects using data. Generally, M&S tool refers to a software that encapsulates a data-driven simulation model within it. Development of a M&S tool usually requires additional steps beyond those shown for simulation model development in figure 2-2.

MS&A tools should be developed for homeland security applications rather than building ad hoc models for each problem. The MS&A tools are essentially data driven models encapsulated in user-friendly software. The process of model development will stay the same albeit based on a problem formulation and setting of objectives comprehending application across identified range of factors. The model would go through a number of validation cycles using multiple test data sets representing scenarios within the intended use of the tool. The model development cycle will need to be integrated within a full software development cycle incorporating features that enable its application by its target consumers for the defined intended use. A potential process for development of M&S tool is shown in Figure 2-3.

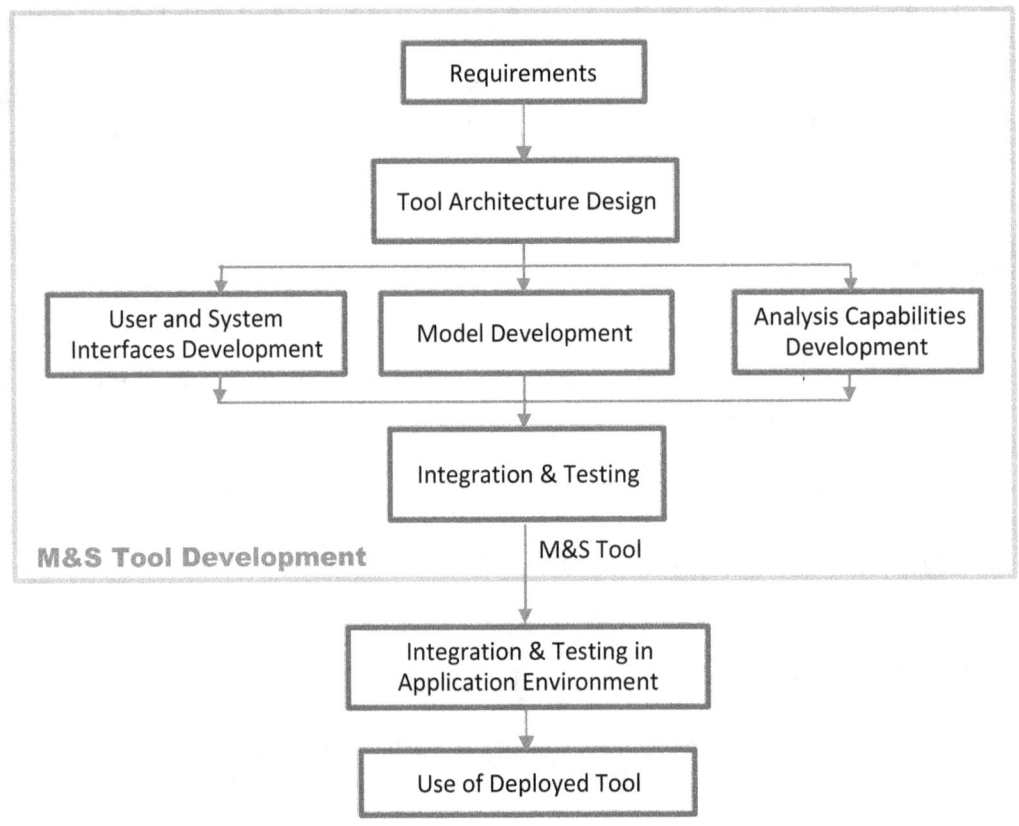

Figure 2-3: A potential process for M&S tool development and deployment

The figure shows a potential process that integrates the simulation model development within the software development cycle. The box with the text "Model Development" in the center of the figure corresponds to "Stage I: Model Development" box in Figure 2-2 and includes all the steps involved. While the figure depicts a sequential process for the ease of presentation, the development process can be iterative.

M&S tools that involve representation of multiple phenomena may be created by integrating existing tools and models that represent one or more of the phenomenon involved. A potential process for development of M&S tools that integrate multiple models is shown in figure 2-4.

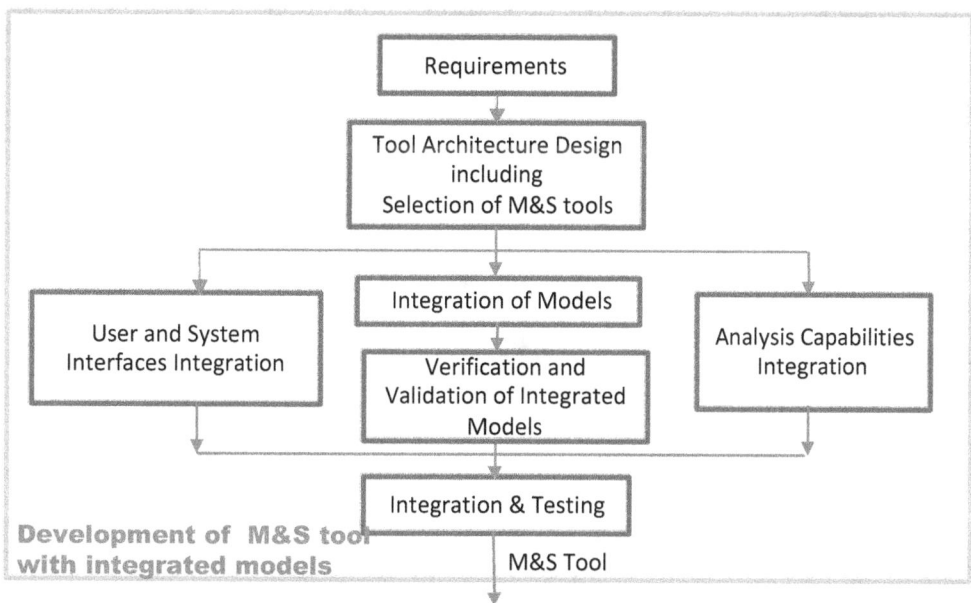

Figure 2-4: A potential process for development of M&S tools with integrated models

As shown in the figure, verification and validation of the integrated models is required. The figure shows a potential process. Such developments have been generally done on ad hoc basis and hence many variants of the process may have been used in practice. The figure shows a tight integration between selected M&S tools as it includes integration of their user interfaces and analysis capability in addition to the model integration. Alternative processes may include integration of the models via exchange of synchronous or asynchronous messages without integrating user interfaces and analysis capabilities. At times, additional models may be developed afresh and integrated with existing models.

It should be noted that while there are a few definitions of recommended process to develop simulation models and there are alternatives defined for software development, there isn't a process defined for development of M&S tool, i.e., software that includes a simulation model. One would imagine that M&S tools would be developed using the structured software development processes. However, Arthur and Nance (2007) report that their findings indicate minimal use of formal software requirements engineering activities within M&S development with the exception of military and government projects. The NSF report on Simulation Based Engineering Science (SBES) points out that "entirely new approaches are needed for the development of the software that will encapsulate the models and methods used in SBES" (NSF 2006, p. 40).

The intent of the figures 2-3 and 2-4 is to provide a context for some of the best practices discussed in this report. The figures thus show potential representations since the development of detailed processes for M&S tool development with single or multiple models is outside the scope.

3 BEST PRACTICES

In this section a set of best practices are defined that are generally applicable across MS&A tools for all homeland security applications, indeed across all applications. The best practices are discussed with respect to homeland security applications where possible in view of the objectives of this report.

A number of best practices are generally applicable, that is, they can be used across all MS&A for homeland security applications. Indeed, the practices discussed in this section are applicable for all MS&A tools and not only those that are relevant for homeland security. The recommended best practices for MS&A are listed below and individually discussed following the list.

- Conceptual modeling practice
- Innovative approaches
- Software engineering practices/ software reliability
- Model confidence/ verification, validation, and accreditation
- Use of standards
- Interoperability
- Performance
- User friendliness and accessibility

The practices above are presented roughly in the sequence that they may be applied during the development and use of simulation software following the steps described in the previous section. The first four practices will primarily impact the model development stage while the latter four practices will impact the use of the model. All of them will need to be addressed during the model development stage. For example, the features for user friendliness and accessibility will have to be built in during the model coding step and will have a major impact on the use of the model.

3.1 Conceptual modeling practice

Practice Introduction

Modeling involves development of a conceptual model to represent the real life system of interest. It requires proper abstraction, that is, identification of potentially important factors in the real life system to incorporate them in the conceptual model. It includes translation of the intended use(s) of the MS&A tool into scope of the model, development of conceptual model(s), and selection of the modeling paradigm. It has been said that modeling is an art and not a science. It may be hence hard to define a detailed guide for the conceptual modeling practice.

It should be noted that the conceptual modeling practice may be considered part of the design phase of the software engineering process. Under the recommended spiral approach, the modeling process may be repeatedly used for defining the scope and detail level of the deliverable for the next development iteration.

There isn't a widely agreed upon definition of a conceptual model particularly with reference to MS&A applications. Lacy et al. (2001) identify two types of conceptual models: a domain-oriented model that provides a detailed representation of the problem domain and a design-oriented model that describes in detail the requirements of the model and is used to design the model code. The former captures the users' point of view while the latter captures the model designer's point of view. The simulations that model physics of the phenomenon may use conceptual models that are quite different from those representing socio-technical systems. The conceptual models in such cases may be representations of involved geometries and physics.

A critical modeling decision is the identification of an appropriate modeling paradigm to represent the phenomena of interest. The two major paradigms are discrete event and continuous simulations with multiple implementation approaches within each such as system dynamics, physics based models, cellular automata, and agent based modeling. The phenomena of interest in the homeland security context include a wide variety such as dispersion of plumes, behavior of population following a major incident, movement and actions of emergency responders, and spread of wild-fire through forests and residential areas. It can be seen that the wide variety of phenomena may require different modeling paradigms for a suitable representation. For example, plume dispersions may be modeled using physics models with differential equations representing associated fluid dynamics in continuous paradigm, behavior of population may be modeled using agent based simulation in discrete event paradigm, movements and actions of emergency responders may be modeled using discrete event simulation, and spread of wild-fire may be modeled using cellular automata in discrete paradigm. The physics based models are typically quite different from systems dynamics models even though both use continuous simulation paradigms.

Modeling paradigm appropriate to the phenomena being modeled and the intended use should be selected for translating the conceptual model to an executable code. In fact, an early selection of the paradigm can also guide the selection of the method for conceptual modeling. For example, causal loop models are typically used to conceptualize the phenomena of interest for system dynamics modeling, one approach for continuous simulation. Process flow charts are typically used to conceptualize processes before developing discrete event models. Entity Relationship diagrams and use cases documented in Unified Modeling Language (UML) may also be employed for conceptual modeling of discrete event simulation models, though such use is not widely reported.

Conceptual models may be validated to the extent possible via model walk-throughs with subject matter experts, customers and end-users. If the conceptual models are based on mathematical equations, they may be validated using existing data on similar systems or a priori knowledge of the phenomenon being modeled[4]. The validation of conceptual models is addressed in the Model confidence/ verification, validation, and accreditation practice presented later in this report.

---

[4] Assessing the correctness of the mathematical approximations in a conceptual model has been termed as "confirmation" by Thacker et al (2004).

Available guidance

There is some general guidance available for the conceptual modeling activity. Robinson (2006) points to the limited availability of specific guidance available but provides an overview of the general guidance available. Three basic approaches are suggested for guiding the process of developing conceptual models. These focus on principles of modeling, methods of simplification, and modeling frameworks.

Pidd (1999) suggests six principles of modeling:
- Model simple; think complicated,
- Be parsimonious; start small and add,
- Divide and conquer; avoid mega models,
- Use metaphors, analogies, and similarities,
- Do not fall in love with data, and,
- Modeling may feel like muddling through.

Zeigler (1976) provides four methods of simplification: dropping unimportant components of the model, using random variables to depict parts of the model, coarsening the range of variables in the model, and grouping components of the model. A number of other authors have provided their simplification methods (Robinson 2006) and can be referred to for additional guidance.

Modeling frameworks provide more specific guidance than the above two basic approaches. Frameworks can include specific steps, templates, or generic model structures for analyzing defined range of problems in identified domains. Guru and Savory (2004) present a set of tables as modeling templates useful for modeling physical security systems.

UML (OMG 2009a) is widely used for developing conceptual models for software design. Conceptual models of business processes have been developed using UML and translated into discrete event simulation (Teilans et al. 2008). Similarly UML models have been used for developing simulations of complex technical systems with software and physical components (Axelsson 2002).

Systems Modeling Languages (SysML; OMG 2009b) was developed through a joint initiative of the International Council on Systems Engineering (INCOSE) and Object Management Group (OMG) Systems Engineering special-interest group. It can be viewed as an extension of UML for systems engineering applications. SysML has been used for conceptual modeling of M&S applications. Huang et al. (2008) report the use of SysML for developing models of wafer fabs. Rao et al. (2008) describe use of SysML for modeling global earth observation system of systems.

Physics-based models may use mathematical equations and schematic diagrams as conceptual models. No de jure standards were identified for conceptual modeling of physics based models. Steinhauser (2008) provides a classification scheme for the physics based models in materials science based on the scale of phenomenon into atomic, micro, meso, and macro. He identifies the use of quantum mechanics and microstructure simulations at the nanoscale and coarse-grained atomistic simulations, and of classical particle approach solving Newton's equations of motion at the meso- and macroscale.

The Real-time Platform Reference Federation Object Model (RPRFOM; SISO 1999) is a reference model that helps in developing conceptual models for distributed simulation models based on the High Level Architecture (HLA; Kuhl et al. 1999).

The Base Object Models (BOMs; SISO 2006) templates provide guidance for output of the conceptual modeling process. They serve as a foundation for the design of executable software code and integration of interoperable simulations. The BOMs include static description of real world in terms of conceptual entities and conceptual events. The BOMs also include interactions of the conceptual entities in terms of patterns of interplay and state machines to represent the corresponding real world relations. Together, the static and dynamic representations can be used as a basis for development of simulation model by the simulation software analysts.

The selection of a modeling paradigm is driven by nature of the problem. If the problem includes modeling of phenomenon that are best represented using continuous time such as spread of plume and flow of water out of a tank, the continuous paradigm would be appropriate. Some of the problems may be modeled either way based on the perspective of the modeler. In general, if events are to be modeled at detailed levels and individual units are to be tracked, discrete event simulation is more appropriate. If the interest is in modeling a phenomenon at high abstraction level, system dynamics models may be more appropriate. There isn't much guidance available regarding selection of one paradigm versus the other. Tako and Robinson (2009) suggest that modeler tend to use the paradigm that they are more familiar with. Using an empirical study they conclude that users do not see much difference between the two based on looking at the outputs.

Recommended Implementation

This practice impacts the "model conceptualization" step in model development process (identified in figure 2-2). Good conceptual modeling practice includes identification and clear communication of the intended use of the model. The intended use and the anticipated contexts should be clearly documented. The intended use should drive the determination of scope of the model. The assumptions made as part of the conceptualization and abstraction of the model should be unambiguously defined and documented.

Developers should utilize the principles of modeling provided by Pidd (1999), the methods of simplification provided by Zeigler (1976), and available modeling frameworks applicable to the problem under study. It is recommended that leading MS&A professionals involved in homeland security domain area develop modeling frameworks for guidance.

The conceptual models should be developed using SysML to enable wider understanding within the modeling and simulation community. If distributed simulation implementation is anticipated, the outputs of the conceptual modeling process should be documented using the Base Object Models and RPRFOM standards defined by Simulation Interoperability Standards Organization

(SISO 1999 and 2006). Admittedly, the SysML may not be well suited for developing conceptual models of physics based models that usually include mathematical equations and diagrams.

The selection of appropriate paradigm for modeling a problem should be determined only after careful consideration and not limited by the background of the modelers. If possible, people familiar with multiple paradigms should be involved in making the decision. Some of the available commercial software (e.g., AnyLogic and GoldSim) allow using a combination of discrete event, agent based, and system dynamics paradigms and should be considered if the problem requires multiple paradigms.

Use for Legacy vs New Applications

Similar to software engineering practice, good conceptual modeling practice should be used for all new applications. Good conceptual modeling practices should also be used to guide any enhancements in legacy applications.

Roles and responsibilities

Use of good modeling process is mostly dependent on the analysts involved in developing the model and the users who help define the desired capabilities of the application. Program and project managers should ensure that the development and documentation of conceptual models is part of the project plan and that it is adhered to. The customers and selected end-users should participate in the validation of the conceptual models to establish a good platform for the implementation of the models.

Costs/benefits

Good conceptual modeling practice can provide several benefits. Balci et al. (2008) recommend conceptual modeling as the primary means to increase reuse among M&S applications and consequent significant economic benefits. Emergency response management problem domain is used as an example to discuss the use and benefits of conceptual modeling. They present a number of benefits of conceptual modeling including assistance in designing M&S applications in the associate problem domain, enabling effective communications in large scale M&S projects, assistance in overcoming complexity of designing in large scale M&S applications, and assistance in verification and validation of the M&S applications.

Metrics

The achievement of this practice can be judged along a continuum going through the following stages:
- Use of conceptual modeling reflected by documented models
- Use of conceptual modeling compliant with identified standards.
- Use of modeling frameworks and conceptual models available for the problem domain.

A potential associated measure is percent reuse of available conceptual models in development of the M&S application.

Practice Conclusion

Use of conceptual modeling before moving to programming the simulation model is a best practice in M&S application development. A standard modeling language such as SysML improves the communication of the conceptual model within and outside the project team. The conceptual model should be translated into simulation code with a paradigm that is appropriate for the problem under study. The adoption of M&S for homeland security applications can be substantially accelerated through the development of modeling frameworks and reusable conceptual models.

3.2 Innovative approaches

Practice Introduction

This practice refers to the use of innovative and unique elements in MS&A tools that enable new applications or improve the possibility of the use of the tools through such features as more accurate output in limited time, improved user experience, and lower requirements for infrastructure than others. An example of use of innovative approach is the RealOpt system developed by researchers at Georgia Tech for planning large-scale emergency dispensing clinics to respond to biological threats and infectious disease outbreaks. The system will be useful for public health administrators for quickly evaluating clinic design and staffing scenario following a bio-agent release or identification of a rapidly spreading disease. The RealOpt system uses innovative combination of heuristics and simulation to reduce the computation time for developing facility layout and staffing recommendation. For a test smallpox scenario, RealOpt provided its recommendations within 1 CPU minute compared to 5-10 hours required by a prototype system built using commercial software (Lee et al. 2006).

The field of computer simulation, particularly for modeling physical phenomenon, has gained from innovative algorithms since its advent in late 1940s. A major step was the development and use of Metropolis algorithm in 1953 that employed Monte Carlo method to calculate the equation of state in two dimensions for a system of rigid spheres (Steinhauser 2008). Two other algorithms, the Fast Fourier Transform developed in 1965 and Fast Multipole algorithm developed in 1987, led to significant reductions in computations for simulation of physical phenomenon. All the three algorithm mentioned above have been identified as among the ten most significant algorithms of 20th century (Dongarra and Sullivan 2000).

The development of innovative algorithms and increasing computing power has opened up opportunities for modeling larger and more complex problems than in the past. Multiple challenges hence continue to be identified as new and more complex problems get targeted. The increasing availability and use of technology has also given rise to the challenge of dealing with vast amounts of data. A large number of tracking systems, sensors, and data logging applications are continuously collecting data for many aspects relevant to homeland security applications.

Simulations themselves can generate volumes of data. There is more data being generated than can be analyzed with current human and computing resources. Another challenge is determining provenance of data, the lineage of data in terms of various key events that occur during its lifecycle and other related information associated with the events. Provenance can help users to share, discover, and reuse data and thus help in collaborative analysis. Provenance is clearly important for intelligence applications associated with homeland security to ensure that quality data is used for analysis and supporting decision making. For MS&A applications, provenance of data can be used to identify the data processed using accredited M&S tools and to differentiate between assumptions and empirical data for improved understanding of results. Overall, a number of challenges exist where innovation can help and in turn can open up opportunities for increased applications for MS&A.

Available guidance

The guidance in the area of innovative approaches comes from academics who have analyzed innovation in industry and proposed theories and frameworks. Anthony et al (2006) credit Professor Clayton Christensen of Harvard Business School for developing the disruptive innovation theory. Professor Christensen suggests looking for new ways to meet and grow the demand of customers who may be looking for lower cost products that provide the basic functionality. Anthony et al. (2006) suggest three ways for organizations to create growth through disruption as listed below.
1. The Back Scratcher: Make it easier and simpler for people to get an important job done.
2. The Extreme Makeover: Find a way to prosper at the low end of established markets by giving people good enough solutions at low prices.
3. The Bottleneck Buster: Expand a market by removing a barrier to consumption.

The three ways above have been proposed for companies competing in commercial environment. However, organizations developing modeling and simulation for homeland security applications can utilize these basic approaches to guide innovation. For example, any innovative approaches that remove the barrier of long computation times can be classified as the bottleneck buster approach.

Goffin and Mitchell (2005) present a pentathlon framework for innovation strategy management. They recommend that organizations work on all the five elements of the pentathlon framework listed below.
1. Creating an innovation strategy
2. Generating ideas
3. Prioritizing and selecting from these
4. Implementing the ideas selected
5. Involving people from across the business.

They suggest that innovation strategy be created considering the stakeholder satisfaction, particularly that of the end users. Again, while the framework has been proposed for commercial environments, it is applicable to the technical area discussed here.

While most available literature on innovation is focused on commercial environments, there is some work focusing on technical areas. Rushby (2006) presents an example of harnessing disruptive innovation for finite state model verification. He presents satisfiability modulo theories (SMT) solver as a disruptive innovation. It uses techniques for theorem proving in ways that enable model checking of computer programs. Paul et al. (2006) implement and evaluate an innovative approach for simulation training acquisitions. The approach proposes acquisition of training as a service acquisition with a private sector "tool vendor" marketplace to support it rather than as an acquisition of training simulators that has been case traditionally. The challenges in the implementation of such an approach are identified. It is recommended that a prototype be developed first and development and implementation plans for the prototype are provided. Smith (2007) identifies game technology as a disruptive innovation in the military simulation training field. Serious games, application of game technology to serious training applications, allow the military to meet the training needs for a large number of personnel distributed all across the globe at a much lower cost than the typical large war simulations they use. There is potential for a similar disruptive application of game technology for homeland security training applications.

Recommended Implementation

Innovative approaches can be used to improve all steps in model development process (shown in figure 2-2) and in M&S tool development process (shown in figures 2-3 and 2-4). The scientific community is occasionally accused of looking for highly sophisticated approaches that may be applicable for few complex cases while the need may be for better ways of satisfying the basic applications for a large number of simple cases. This indeed appears to be the premise of disruptive innovations. It is recommended that the developers and implementers of homeland security applications employing modeling and simulation focus on identifying innovative ways to meet the needs of the involved organizations and end users. The developers and implementers should identify the promising approach among the three suggested by Anthony et al. (2006) and listed above as they embark on identifying new applications. Their home organizations should create a supportive environment for innovation using a framework such as the pentathlon mentioned above.

Use for Legacy vs New Applications

Innovative ways can be employed to use legacy applications in new ways or for modifying them using new approaches. The modification of legacy applications using new approaches may be almost as demanding as developing a new one and coupled with the availability of new technology, it may not be a worthwhile endeavor. It would be better to develop a new application that uses the innovative approach and builds on the latest technology.

New MS&A applications can definitely gain from use of innovative approaches. Developers should question each part of the design of a new MS&A application to ensure that the best possible approach is being used rather than simple rewrite of traditional approaches using new software and hardware technologies.

Roles and responsibilities

The primary opportunity for introducing innovation in development and implementation of modeling and simulation for homeland security applications is available to the modelers and programmers. They should identify innovative approaches for modeling the phenomenon being addressed and for deploying it for use. All aspects of modeling and its translation to executable code need to be considered for improvement including the math, computer code, data, and analysis procedures. The math needs to be well understood before innovating. Innovative approaches should be especially considered for reducing computations as exemplified in the case of RealOpt mentioned above. They also need to consider the inputs and outputs of the simulations and to develop innovative ways to minimize the time required for providing the inputs and the fully understand the outputs.

While the primary opportunity is available to developers, innovation is not limited to that role only as clear from the work of Paul et al. (2006) mentioned above that looked at innovative business models for simulation training acquisitions. Everyone intimately involved with an application, including program managers and end users, should be looking out for improving all aspects of modeling and simulation for homeland security applications. The premise of disruptive innovations, i.e., that of looking for new ways of doing things that impact a much larger number of end users with basic requirements, should be considered in identifying program and project objectives.

Costs/benefits

No documented cost benefit analysis was found on use of innovative approaches in MS&A applications and hence some related studies are considered for indication. Homeland security related applications of MS&A will typically be totally or partially funded by public sources. There may be some instances of such applications in commercial arena such as risk analysis for insurance purposes. Powell (2006) discusses a standard benefit-cost methodology for publicly funded science and technology programs. It is suggested that both social returns (return to nation and project participants on total investment from all sources) and public returns (returns to nation attributable to the public funding) be computed. The report quotes results from earlier studies indicating private rate of returns averaging 25%-36% and social rate of returns averaging 50%-70%. A study commissioned by DoD on effectiveness of M&S in weapons systems acquisition process found that M&S reduced risk throughout development cycles, improved system performance, and lowered total life cycle costs. Most programs studied cited cost avoidance or resource conservation as benefits (Patenaude 1996). A National Research Council report estimates a return of investment of 25:1 through use of M&S in small acquisition programs with no more $20M invested in simulation (NRC 1997). Gordon (2000) presents a collection of anecdotal evidence for benefits from simulation for defense applications across war gaming, experimentation, assessment, acquisition, evaluation, training, and decision support to combat operations. He suggests that the benefit of M&S should be evaluated in terms of the combat readiness gained rather than cost savings.

Metrics

The metrics for use of innovative approaches are generally defined for commercial environments and are related to investment in research and development and number of new products. Dehoff et al. (2007) propose return on innovation investment (Internal Rate of Return for each innovation project against annual project expenditure) and freshness index (revenue from products launched in last 3 years against total annual revenue). Chan et al. (2008) provide a list of thirteen metrics identified through a survey of companies. The metrics that may be applied to a non-commercial environment include:

- Customer (end-user) satisfaction with new products or services,
- Numbers of ideas and concept in the pipeline, and,
- Number of new products and services launched.

The above may apply to an organization involved in research and development of MS&A for homeland security applications. A similar list is proposed below for innovation metrics for an application or a group of related applications.

- Customer (end-user) satisfaction with new features in application
- Numbers of ideas and concept for new features in the pipeline
- Number of new features released
- Number of new M&S tools put in use
- Number of new homeland security applications based on MS&A

Practice Conclusion

Organizations involved in development of MS&A for homeland security applications should continually search for innovative features and capabilities that may help the end-users in harnessing the power of the MS&A tools for any of the application types. They should also take advantage of evolving computer and networking technology for application of MS&A capabilities to new problems. The lessons learned from previous implementations should be used to identify the needs for innovative approaches.

Use of innovative approaches may help in any or multiple of a number of aspects including developing a better conceptual model, populating it with data, executing the simulation faster, analyzing the outputs, and disseminating the results to the decision makers. In general, all development organizations do continually look for the best solutions to problems at hand. However, specific attention to policies, procedures, and organization support to encourage innovation may create a supportive environment for increased innovative activity.

3.3 Software engineering practices/ software reliability

Practice Introduction

Software engineering practices refer to the approach used to design and develop modeling and simulation software for the purpose of this report. Clearly this practice would apply only when a new MS&A based application needs to be developed. In some cases, analysts may be able to use a commercial or government off the shelf tool for an application. In such cases, the use of software engineering may apply only to any interfaces and/or middleware that may be needed for access by the decision maker, for accessing data, and for interacting with other applications.

Use of a systematic approach to develop and test the software increases the confidence that it can be executed to generate the results, i.e., improves the reliability of the software and repeatability of results. Use of good software engineering practices makes the task of development well organized and the generated code itself well organized, well documented, and easy to maintain. A formal definition of software engineering is provided below.

Software engineering is the application of a systematic, disciplined, quantifiable approach to the development, operation, and maintenance of software, and the study of these approaches; that is, the application of engineering to software (IEEE 2004).

MS&A tools are a specialized kind of software and hence should be created using mature software engineering practices. Nance and Arthur (2006) strongly recommend a focus on software requirements engineering in simulation model development. Use of such practices will ensure that the MS&A tools have the required capabilities and it is reliable among other benefits.

Available guidance

The software development processes used for developing MS&A tools should qualify among higher levels of the Capability Maturity Model Integration (CMMI) of the Software Engineering Institute (SEI). CMMI provides a process improvement approach for improving the maturity of the software development process using a structured approach and elements. It is based on earlier standards including Capability Maturity Model (CMM) from SEI and EIA-731 Systems Engineering. CMMI describes five distinct levels of maturity (Royce 2002):
- Level 1 – initial – unpredictable results,
- Level 2 – managed – repeatable project performance,
- Level 3 – defined – improving project performance within an organization,
- Level 4 – quantitatively managed – improving organizational performance, and,
- Level 5 – optimized – rapidly reconfigurable organizational performance as well as quantitative, continuous process improvement.

A number of guidelines have been developed over the years for software development process. A good source is the *Guide to Software Engineering Body of Knowledge* (IEEE 2004) developed under the auspices of IEEE Computer Society. The IEEE/EIA 12207 standard for software lifecycle processes subsumed earlier standards including ISO 12207 and J-STD-016, which in turn superseded MIL-STD-498 that superseded DoD-STD-2167 and DoD-STD-7935A. The

IEEE/EIA 12207 defines software lifecycle processes including its industry implementation, life cycle data, and implementation considerations (IEEE 1996 and 1997). It has been adopted by the US Department of Defense (DoD) as a software development standard. The standard defines processes in three categories, primary life cycle processes, supporting life cycle processes, and organizational life cycle processes. It should be noted that the supporting life cycle processes include verification and validation processes that refer to qualification testing of the software. The verification and validation processes for MS&A tools are specialized and hence addressed separately in the next best practice.

The whole development process should be driven by user requirements. The set of requirements should be developed based on clearly identified and documented intended uses of the system. The intended use documentation may at time be variously referred to as system specification, user requirements, or functional requirements. Balci (2004) recommends use of a tool like Evaluation Environment for collaborative application of a quality model and emphasizes the need for rigorous quality assessment of M&S products including M&S requirement specification.

The overall software development process may follow an established model such as the waterfall, prototype, or spiral. The latter two may also be seen as part of the agile software development model. The waterfall development model is a linear development process with detailed planning for the full length of the project, while the agile approach is a combination of linear and iterative development processes with detailed planning for only the next increment. In case of agile model, each increment should follow a structured development cycle from requirements to testing. Agile approaches are recommended if the development team has close access to users, else a waterfall model may be more appropriate. Also, a CMMI based approach may be suitable for large multi-team multi-site long living simulation projects while agile approaches may be suitable for small simulation projects (Sargent et al. 2006).

Few simulation development projects focus on requirements with the notable exception of those developed by military and government agencies (Nance and Arthur 2006). Two notable examples of emphasis on requirements in simulation development are identified. The first one is on waste management (Miller et al. 2003) while the second one is on Global Systems Simulation Program primarily using continuous system simulation (Drake et al. 2009).

Oberkampf et al. (2007) suggest that CMMI approach is not suitable for M&S applications and present a predictive capability maturity model (PCMM) that should be used to ensure the maturity of the computation simulation process elements. While the elements are suitable for and are indeed recommended for assessment of practices later in this document, the overall PCMM model does not appear to assess and promote a requirements driven software engineering practice as is the case with CMMI or the spiral approach.

Recommended Implementation

The use of good software engineering practice should extend throughout the development process of the MS&A tool (shown in figures 2-3 and 2-4). The proposal for development of a MS&A tool should identify the CMMI maturity level of the proposing organization and the defined plans in the proposal should reflect the use of good software engineering practice. An exception may be made for academic organizations if the MS&A tool development involves substantial research issues. With the wide range of phenomenon that can be addressed for homeland security applications of MS&A, there may be a number of research issues to be tackled. It is recommended though that the development of the research version of the tool should also follow structured approaches and a partner organization should be involved for transitioning to production version of the tool. A spiral approach may be more suited for development of the research version of the tool.

The MS&A tool development effort should go through a full software development cycle including planning, requirements analysis, design, coding, unit testing, and acceptance testing. The process should include good software configuration management, development of information models, and detailed documentation. The testing steps may overlap with verification, validation, and accreditation, part of the best practice devoted to that topic.

In general it is recommended that the MS&A tools for homeland security applications be developed for usage across a number of jurisdictions. The development process for such an environment is generally recommended to be CMMI based. Developing applications for potential multi-site applications would require a large effort to gather user requirements. However, such efforts can be quite time consuming. A hybrid approach is recommended where a prototype should be developed using a spiral approach working with a select small group of users followed by development of the version for deployment across multiple sites using a CMMI based approach. Availability of a prototype is expected to significantly accelerate the process of generating requirements from the users at multiple sites.

An iterative approach is recommended for development of the prototype versions of MS&A tools to ensure that the development meets the needs of the users. Iterative approaches include spiral and agile methodologies such as extreme programming. Nearly all agile methodologies use some variation on spiral development (Kussmaul 2005). The emergency managers and first responders are the primary users of the outputs of MS&A tools for planning and operations applications. These primary users in general may not have a good understanding of MS&A tools. Use of an iterative approach will allow educating the selected primary users on the capabilities of MS&A tool and have them adjust their specified requirements and generate appropriate additional requirements through successive iterations. The underlying assumption here is that the developers have a close access to the primary users for development of the prototypes. The MS&A tools development process may follow the CMMI based approach all through if this assumption does not hold. The key idea is that a structured requirements-driven approach should be followed for development of MS&A tools. The specialized nature of MS&A tools should not translate into informal unstructured development.

Use for Legacy vs New Applications

Good software engineering practices should be used for development of all new MS&A applications. They should also be used for any modifications and enhancements to legacy MS&A applications. If the legacy application is quite outdated and is very hard to modify due to original development being unstructured, redeveloping the application using good software engineering practices and the latest technologies should be seriously considered.

Development of new applications should follow the recommended hybrid approach. However, for modifications or redevelopment of legacy applications a CMMI based approach should be used since the legacy application can serve as the prototype to accelerate the generation of additional requirements by multi-site multi-team users.

Roles and responsibilities

Developers of MS&A tools for homeland security applications have the primary responsibility to utilize good software engineering practices. The developers may include national laboratories, DHS university centers of excellence, and commercial organizations.

The end users of the MS&A tools need to stay closely involved in providing and updating requirements and feedback on the successively developed versions under the spiral approach. The users need to be deeply involved in defining the requirements if a waterfall approach is used. The end users may include personnel from federal, state, or local organizations executing homeland security functions including emergency response, border protection, aviation security, etc. depending on the scope of the MS&A tool.

The contracting program manager would need to ensure that the contractor organization follows software engineering processes that qualify for higher maturity levels of CMMI and that the submitted project plans and other documentation reflect the use of a structured approach.

Costs/benefits

Use of good software engineering practices results in higher productivity, higher quality software, less maintenance, and faster cycle times among other benefits. Rico (2002) calculated a benefit to cost ratio of 11:1 and a return on investment of 1,044% through use of CMMI. It should be noted that Rico (2002) indicated that other methods for software process improvement may provide higher returns on investment. Galin and Avrahami (2006) considered 19 studies to calculate benefits of CMM programs. They found mean improvements of 48% in error density, 52% in productivity, 39% in rework, 38% in cycle time, 45% in schedule fidelity, 63% in error detection effectiveness, and 360% in ROI.

Metrics

Pursuant to use of CMMI as the guidance for improving the software engineering practices, the associated maturity levels should be used to identify the current practice and to set targets for improvement. In addition, similar to the study by Galin and Avrahami (2006), performance measures for the MS&A software development effort should include error density, productivity, rework, cycle time, schedule fidelity, error detection effectiveness, and ROI.

Practice Conclusion

The best practice for MS&A tool development is to follow a structured software engineering process that is driven by requirements. The actual guidelines used (such as CMMI, or IEEE/ EIA 12207) would depend on the developing and contracting organization, but it should follow a disciplined focus on requirements and a structured process to build and deliver to them. It is recommended that a hybrid approach be used combining an iterative approach for the development of the initial prototype and a CMMI based process for the version for deployment to users (generally referred to as the production version in software industry).

3.4 Model confidence/ Verification, validation and accreditation procedures

Practice Introduction

M&S applications are of little value if there is not a high degree of confidence in their results. Homeland security decisions based on models and simulations must be reliable, they may involve national security, loss of human life, and/or large expenditures of public funds and other resources. It is critical that a model or a simulation and associated data are correct. Verification, validation, and accreditation (VV&A) procedures are the mechanisms that are typically used to assure quality in modeling and simulation. VV&A helps ensure the reliability of simulation models and data for a specific intended use. The U.S. Department of Defense has invested considerable resources in the establishment of VV&A procedures.

VV&A terminology has been defined by the U.S. Department of Defense as follows: *Verification* is the process of determining that a model or simulation implementation and its associated data accurately represent the developer's conceptual description and specifications (DoD 2009). *Validation* is the process of determining the degree to which a model or simulation and its associated data are an accurate representation of the real world from the perspective of the intended uses of the model (DoD 2009). *Accreditation* is the official certification that a model or simulation and its associated data are acceptable for use for a specific purpose (DoD 2009). Accreditation is conferred by the organization best positioned to make the judgment that the model or simulation in question is acceptable. An accrediting organization may be an operational user, a program office, or a contractor, depending upon the purpose of the model or simulation.

The M&S tools used for homeland security applications should go through structured verification, validation and accreditation (VV&A) procedures. It is understood that DHS is developing formal VV&A policies. Until the policies are developed, it is recommended that M&S tools for homeland security applications go through a structured verification and validation (V&V) approach. Some of the available structured V&V procedures are discussed in this section. Formal accreditation should be carried out once the official DHS policy for VV&A is available.

The V&V process should utilize a set of validated data. The U.S. Department of Defense (DoD) learned over time that V&V of a model cannot be decoupled from validation of data (Blake, 2008). A best practice should include validation of data as part of the VV&A procedures.

Available guidance

The V&V process should be driven by the same requirements as used for the development of the MS&A tool. The V&V plan should be based on the prioritization of different phenomena modeled by the particular MS&A tool. Phenomena Identification and Ranking Table (PIRT) can be used for prioritizing the V&V efforts including the development of test data for physics based M&S tools (Trucano and Moya, 1999). PIRT involves definition of the primary drivers for the use of the modeling tool, and development of a table incorporating the phenomena modeled, their importance, and the adequacy of the modeling with respect to the intended use. More effort should be spent on V&V of the tools with respect to the phenomena that are evaluated to be at high priority through PIRT. Similarly, test data sets should be focused on testing the tool's capability for modeling the high priority phenomena.

Over 100 V&V techniques exist (Balci 2004). V&V techniques applicable in different stages of M&S application development lifecycle have been identified (Balci 2003). Sargent (2010) provides a recommended V&V procedure. A number of guidance documents are also available. The U.S. Department of Navy developed a handbook for VV&A (DON 2004). An IEEE standard provides recommended practice for VV&A of a federation of distributed simulations integrated using the High Level Architecture (IEEE 2007). A product development group at Simulation Interoperability Standards Organization (SISO) is developing a generic methodology for verification, validation and acceptance for models, simulations and data (SISO, 2010a). Domain specific guidance is available for V&V of computational fluid dynamics simulations (AIAA 1998) and computational solid mechanics (ASME 2006).

The standard for modeling and simulation from National Aeronautics and Space Administration (NASA 2008) has the overall goal "to ensure that the credibility of the results from M&S is properly conveyed to those making critical decisions." Thus, while it covers a number of aspects related to use of modeling and simulation, one of the primary focus is on the credibility of the M&S results and hence on V&V. The standard defines three requirements each for verification and validation calling for documentation of the processes used and the results. A credibility assessment scale is defined that uses three categories that together include eight factors as follows: M&S Development (Verification, Validation); M&S Operations (Input Pedigree, Results Uncertainty, Results Robustness); and Supporting Evidence (Use History, M&S Management, People Qualifications). A five-level assessment of credibility is defined for each factor. M&S results can be assessed for each factor and the scores rolled up into a single number that represents the summary credibility assessment.

DoD has a VV&A policy (DoD 2009) as well as significant guidance available on-line in the form of VV&A recommended practices guide (RPG; DoD 2006). The VV&A RPG provides guidance across a wide range of topics that allows a user to select the applicable sections. For example, it provides guidance for validating new models, legacy models, and simulation data. The RPG is a good source for VV&A guidance.

The predictive capability maturity model (PCMM) presented by Oberkampf et al. (2007) assesses the maturity of the computation simulation process elements. With the emphasis in PCMM on assessment and reviews, user should have a higher confidence in the MS&A applications identified at higher levels of maturity. The PCMM is hence seen more relevant to the practice related to model confidence rather than software engineering practice discussed earlier.

The PCMM identifies six model elements that may be assessed on four levels of maturity. The six model elements are: representation and geometric fidelity, physics and material model facility, code verification, solution verification, model validation, and uncertainty quantification and sensitivity analysis. The four maturity levels are defined in general as below:
- Level 0
    - Little or no assessment of completeness and characterization
    - Individual judgment and experience
- Level 1
    - Some informal assessment of completeness and characterization
    - Some evidence of maturity
- Level 2
    - Some formal assessment of completeness and characterization
    - Significant evidence of maturity
    - Some assessments have been made by internal peer review
- Level 3
    - Formal assessment of completeness and characterization
    - Detailed and complete evidence of maturity
    - Essentially all assessments have been made by independent peer review

Oberkampf et al (2007) caution that the maturity of the process is not necessarily the same as predictive accuracy or the predictive adequacy required for a particular project. The PCMM assesses the achievement on identified elements and provides suggestions on composite assessment.

Recommended Implementation

VV&A practice will impact multiple aspects in model development and M&S tool development. It is clearly implemented in the verification and validation steps in the model development process (shown in figure 2-2). In addition, it will impact the data collection process since the data needs to be validated, and it will impact the model conceptualization step as conceptual models should be validated to the extent possible. The M&S tool development process (show in figure 2-4) calls for use of V&V of integrated models if multiple models are utilized to address the problem.

The DoD VV&A recommended process guide (RPG) should be used for M&S applications for homeland security until guidance for such purpose is developed by DHS. It is recognized that while the verification and validation (V&V) steps in the RPG can be followed for homeland security M&S applications, accreditation cannot be carried out until DHS identifies and authorizes agents for the purpose.

Briefly, the basic activities in the V&V process defined in RPG are as below (extracted from DoD 2006):
- Verify M&S Requirements – confirming that the requirements for the simulation match those needed for the current problem, and are correct, consistent, clear, and complete.
- Develop V&V Plan – identifying the objectives, priorities, tasks, and products of the V&V effort; establishing schedules; allocating resources; etc. in coordination with simulation development and accreditation plans.
- Validate Conceptual Model – confirming that the capabilities indicated in the conceptual model embody all the capabilities necessary to meet the requirements.
- Verify Design – determining that the design is faithful to the conceptual model, and contains all the elements necessary to provide all needed capabilities without adding unneeded capabilities.
- Verify Implementation – determining that the code is correct and is implemented correctly on the hardware.
- Validate Results – determining the extent to which the simulation addresses the requirements of the intended use.

Each of the above basic activities also includes specific activities contributing to data validation.

The accreditation follows the completion of the integrated M&S development and validation process. It includes the following steps (extracted from DoD 2006):

- Develop Accreditation Plan – the accreditation plan should identify all the information needed to perform the accreditation assessment and their priorities, tasks, schedules, participants, etc., in coordination with simulation development and V&V plans.
- Collect and Evaluate Accreditation Information – the information needed for the assessment is collected from the V&V effort and other sources and evaluated to determine its completeness.
- Perform Accreditation Assessment – the fitness of the simulation is assessed using all the evidence collected from the V&V effort and other sources, and an accreditation report and recommendations are prepared for the user.

The RPG should be referred to for more details on VV&A process.

Use for Legacy vs New Applications

Both new and legacy M&S applications should go through a VV&A process. In both cases, the VV&A process is integrated with the development process (development of completed model for new applications and of enhancements or modifications for the legacy application). The basic activities of V&V process defined above should be executed for new M&S applications. The extent of V&V for legacy applications depends on the amount of modification required for reuse. Major modifications, i.e., replacing or adding more than 30% of the code (DoD 2006), should require the complete modified model to go through the V&V process as defined above. Minor modifications should require executing the basic V&V activities only on the parts added or modified.

Roles and responsibilities

Dedicated resources, identified as V&V agents, are recommended for executing the V&V process. The resources may be part of the M&S development organization or provided by an independent third party. The V&V agents should lead the basic V&V activities defined earlier. The users and the developers of M&S applications should be involved in the V&V activities in approve and assist roles respectively. The V&V agent should also receive guidance from the identified accreditation agent for the development. Figure 3-1 shows the roles and responsibilities as defined in the DoD VV&A recommended process guide (RPG; DoD 2006).

Typical Roles and Responsibilities Associated with M&S VV&A							
Role / Activity	User	M&S PM	Developer	V&V Agent	Accreditation Agent	SME
Define Requirements	Lead / Approve	Monitor	Assist	Review	Review	Assist
Define Measures	Lead / Approve	Monitor	Assist	Assist	Assist	Assist
Define Acceptability Criteria	Assist / Approve	Monitor	Assist	Assist	Lead	Assist
Plan M&S Development or Modification[1]	Assist* / Lead / Approve	Lead*	Assist	Assist	Assist	
Develop V&V Plans	Review	Assist / Approve	Review	Lead	Assist	
Develop Accreditation Plan	Review / Approve	Assist		Assist	Lead	
Verify Requirements	Lead** / Approve	Monitor	Assist	Lead** (primary)	Assist	Assist
Develop Conceptual Model[2]	Assist / Approve	Monitor	Lead			Assist
Validate Conceptual Model	Assist / Approve	Monitor	Assist	Lead		Assist
Develop Design[3]		Monitor / Approve	Perform			
Verify Design	Approve	Monitor	Assist	Lead		Assist
Implement Design		Monitor / Approve	Perform			
Verify & Validate Data	Approve	Monitor	Assist	Lead		Perform
Verify Implementation (Code)	Approve	Monitor	Assist	Lead		Assist
Test Implementation	Approve	Monitor	Lead	Assist		Assist
Validate Results	Assist / Approve	Monitor	Assist	Lead		Assist
Prepare V&V Report				Perform		
Configure for Use	Assist* / Lead / Approve	Lead*	Assist	Assist		
Gather Additional Accreditation Info	Monitor	Assist		Assist	Lead	Assist
Conduct Accreditation Assessment	Monitor				Perform	Assist
Lead	Leads the task. Normally involves active participation from others					
Assist	Actively participates in task (e.g., conducting tests, providing information)					
Approve	Determines when an activity is satisfactorily completed and another can begin. Determines what activity should be pursued next (e.g., whether to continue to the next scheduled activity or return to a previous activity).					

*In general, this activity is led by the MS PM in new M&S developments and by the User in the modification of a legacy simulation.
**This activity is led by the V&V Agent when available and by the User when the V&V Agent is not available at the beginning of the effort.
[1] This activity refers to planning and scheduling of any M&S development, modification, or preparation
[2] This activity refers to development of new as well as modification of existing conceptual models
[3] This activity refers to development of new M&S designs as well as modification of existing M&S designs

Figure 3-1: Roles and responsibilities for M&S VV&A (adapted from DoD 2006).

In the DHS context, the contracting program manager and project managers should ensure that the development plan includes the basic V&V activities defined earlier and that these activities are executed to the satisfaction of the users and subject matters experts. Ideally the contracting program manager should ensure availability of budget for V&V activities.

An accreditation agent from the identified authority should perform the accreditation with involvement from the users.

Costs/benefits

VV&A process is seen as a necessity and hence should not need to be justified on the basis of cost-benefit analysis. It provides confidence in the results of the model and thus enables decision making based on the outputs of the M&S applications. The benefits of VV&A accrue to the benefits of MS&A and include (DoD 2006):

- Enhanced simulation development process with little, if any, additional cost.
- Reduced overall net simulation development costs.
- Reduced risks and costs of making incorrect program decisions.
- Elimination of up-front, all-or-nothing, go/no-go decision.

DoD focuses on the benefit of reducing development and operational risks of M&S through V&V. The DoD RPG (DoD 2006) accepts that it is rarely economical to uncover and correct all potential defects in simulations through V&V. It identifies the objective of V&V as primarily to minimize the risk that the simulation will produce inaccurate results in a given application by finding all the critical defects. The benefit of V&V, hence, may be estimated in terms of reduced risks. As a corollary, there is a point beyond which the diminishing returns would not justify additional V&V effort. Feather (2004) describes the use of cost benefit trade space at NASA to guide the effort to increase reliability of complex systems. A similar approach may be used to ensure that validation effort is used in a cost effective manner.

Metrics

The metrics for the achievement of this practice should focus on the maturity of the VV&A process. The discussion on measures in the DoD RPG focuses on measures for assessing how well the simulation is able to address the associated M&S requirement, that is, on the last one of the basic activities listed above. It is proposed that maturity of the following basic activities in the DoD RPG be assessed and reported as metrics for achievement of the VV&A practice.

- Validate Conceptual Model
- Verify Design
- Verify Implementation
- Validate Results

The PCMM scheme of four maturity levels can be used to assess the maturity for each of the above basic activities. The maturity of the conceptual model validation for all MS&A

applications other than physics based models can be assessed using the general levels identified. For physics based models, the conceptual model can be viewed as comprising of the first two elements of PCMM, namely, representation and geometric fidelity, and physics and material model facility, and associated maturity levels used. The verify design, verify implementation, and validate results basic activities can be assessed using the maturity levels defined for the PCMM elements code verification, solution verification, and model verification respectively.

The assessment of the validity of the model is recommended to be done using uncertainty quantifications where a referent is available. The referent could be an experimental model or a very similar system to the one being modeled. Attempt should be made to quantify both the irreducible uncertainty (aleatory uncertainty) and reducible uncertainty (epistemic uncertainty) (Thacker et al. 2004). Roy and Oberkampf (2010) provide a framework for characterizing uncertainties from various sources including aleatory, epistemic, model form, and numerical approximation, and define an area validation metric.

Practice Conclusion

A structured VV&A process as part of the MS&A tool development life cycle is a best practice. Execution of a structured VV&A process provides increased confidence in MS&A application outputs and support for their use for decision making. The DoD VV&A RPG should be used as guidance for V&V for MS&A tools and associated data for homeland security application until DHS issues its own VV&A policies and procedures.

3.5 Use of Standards

Practice Introduction

The use of standards for MS&A for a homeland security application, or for that matter, any Information Technology (IT) application, allows it to be employed for other similar purposes at a much lower cost than one that doesn't comply with standards. An excellent application that doesn't conform to standards may be less useful to the enterprise than a good application that does since the standards enable it to be rapidly deployed at multiple sites. The use of standards hence is an important facet of best practice. The previous best practices referred to standards and widely accepted guidance for software engineering, conceptual modeling practice and V&V processes. The best practice discussed in this section concerns the compliance of the MS&A tools with applicable standards not covered in other practices discussed in this document. In general, a tool that uses standards will be easier to integrate with other tools and systems that also comply with the same standards. However, in some cases a tool can be integrated with other commonly used tools and systems (de facto standards) and provide similar benefits without actually complying with de jure standards.

Available guidance

Standards help the homeland security community make more effective and efficient use of MS&A applications. The standards must support the design, development, and implementation of the MS&A applications. Examples of major categories of standards that are relevant to MS&A applications include *Architectures, General Purpose Integration Interfaces, Domain-specific Integration Interfaces, Equipment Specifications, Operational Guidelines, Document Formats* and *Data* (McLean et al. 2008).

Architectures support the overall design or structure of a system or system environment and interactions within a system of systems. Example includes the High Level Architecture that defines the interaction between distributed simulations, each representing a system. Integration interface standards facilitate the interoperation or data exchange between systems. *General Purpose Integration Interfaces* are used to integrate a wide variety of computer applications and are not specific to homeland security or related mission areas. Example interfaces include markup languages, image file formats, and database query languages. *Domain-specific Integration Interfaces* are specific to homeland security related areas, e.g., emergency communications message formats. *Equipment Specifications* define required capabilities, functional characteristics, or rules that ensure quality, safety, and health of users. *Operational Guidelines* define organizational structures, policies, procedures, and protocols. *Document Formats* specify layout and structure for documents in word processing, database, spreadsheet, graphic, presentation, printed, and encoded formats.

The *Data* standards can be classified by the associated major data elements that are relevant for MS&A for homeland security applications. Jain et al (2007) identify thirteen major relevant data elements, namely, *Areas, Building-Structures, Chronology, Demographics, Environment, Hazard-Effects, Incident-Event, Infrastructure-Systems, Organizations, Policies-Procedures-and-Protocols, Response-Operations, Response-Resources,* and *Social-Behaviors.* Applicable standards are identified for each data elements where available, for example, the *Areas* data can be defined using Content Standard for Digital Geospatial Metadata or Governmental Unit Boundary Exchange Standard among others.

Recommended Implementation

Use of standards will improve multiple steps in model development and M&S tool development. For example, use of standards for capturing data relevant to a model will substantially facilitate the data collection step in model development (shown in figure 2-2). Similarly use of a standard representation of the conceptual model will facilitate the model translation step. In the M&S tool development process (shown in figures 2-3 and 2-4), use of standards will facilitate the integration of tool components, integration of multiple models, and the integration of the M&S tool in the application environment among other benefits.

Overlapping competing standards make it difficult for developers of MS&A applications to select the ones that the applications should comply with. The key idea is to ensure that the applications are compliant with relevant de jure and de facto standards. Ideally one would want the applications to comply with all applicable standards, but that would require prohibitive development costs. One should identify the standards being used by other applications that the new MS&A application will interoperate with to guide the identification of relevant standards. Error in choice of standards is a lower risk option than not choosing any and developing proprietary interfaces and operating practices. Sriram et al. (2009) propose an approach for harmonizing information standards across product design and business domains. A similar approach may be used for harmonizing standards in the homeland security modeling and simulation domain.

In some cases, DHS may have identified or sponsored development of integration mechanisms and applications that may integrate with MS&A applications. Such mechanisms and applications should be considered when identifying applicable standards. DHS Science & Technology (S&T) directorate has sponsored the development of Unified Incident Command and Decision Support (UICDS), a middleware that enables modular and scalable information sharing to meet the command and coordination needs for emergency response operations. MS&A applications of the planning and operations type should comprehend the availability of data and associated formats through UICDS. UICDS itself uses a number of standard formats for data inputs and outputs.

Use for Legacy vs New Applications

Both legacy and new applications need to comply with standards to allow their wider implementations. While new applications should be designed with interfaces that comply with applicable standards, legacy applications may require add on interfaces that allow such compliance. The add-on interfaces may read information in standard formats but provide it to the application in its native proprietary formats. Developers have to continually monitor the applicable standards to allow updating the interfaces corresponding to revisions and updates in standards.

Roles and responsibilities

The primary responsibility of ensuring compliance with standards is with the developers of the MS&A applications. Potential users and subject matter experts should help in identifying the applicable standards through clear specification of other applications and systems that the MS&A application will need to interoperate with during its intended use. Program and project managers for MS&A application development should ensure that the plans include identification of relevant standards and associated activities to ensure compliance. The plan should also include tracking of relevant new standards development efforts to ensure any available information is considered and to provide input to such efforts.

Costs/benefits

Use of standards leads to cost reductions in the long term. While no cost benefit studies were located for the use of standards for homeland security MS&A applications, other studies in the past in other areas have estimated significant benefits through use of standards or significant costs due to lack of standards infrastructure. Gallaher et al. (2002) estimated a benefit-to-cost ratio of 11.4 through the use of Standard for the Exchange of Product model data (STEP) in transportation equipment industries. White et al. (2004) estimate the total annual costs of inadequacies in supply chain infrastructures, including critical standards, to be in excess of $5 billion for the automotive industry, and almost $3.9 billion for the electronics industry.

Metrics

Implementation of the standards practice may be assessed using the level of compliance (high, medium, low) achieved by MS&A applications with respect to applicable standards. Only a few areas may have clearly accepted standards that the developers of MS&A applications can design to. It is recommended that DHS provide direction on applicable set of standards using a classification scheme similar to the one discussed above in available guidance. Until such direction is available, program managers should encourage developers to design the applications such that they comply with all the applicable standards and include reporting on standards used and the associated compliance levels.

Practice Conclusion

The use of standards enables wider use of the MS&A tools and hence is a recommended best practice. At times there are multiple standards for the same purpose. In such a case, one has to select the standard that is used more often than others in the specific simulation domain. It is important though to use a standard rather than using a custom practice or development for the purpose at hand. Developing MS&A applications such that they comply with all applicable standards will allow their wider and faster deployment and can generate substantial savings through increased reuse of such applications.

3.6 Interoperability

Practice Introduction

Homeland security applications cover a wide range of scenarios including man-made and natural disasters. A monolithic model to cover the wide range of scenarios is infeasible and would not be desirable for multiple reasons even it were feasible. Customized models for each scenario for each jurisdiction would be a highly inefficient way to use M&S. The most efficient approach is to develop generic data driven component models that can be integrated in combinations required to represent a scenario of interest for a jurisdiction. This approach requires that component models be interoperable.

Interoperability among component models can be established from a low level focused on network connectivity, generally defined as integretability, to a high level focused on alignment of conceptual models, generally defined as composability. Turnitsa (2005) provides a seven level model for conceptual interoperability as follows.

        Level 0: No interoperability
        Level 1: Technical interoperability
        Level 2: Syntactic interoperability
        Level 3: Semantic interoperability
        Level 4: Pragmatic interoperability
        Level 5: Dynamic interoperability
        Level 6: Conceptual interoperability

At the highest level of conceptual interoperability, the assumptions, constraints, and the abstracted conceptual models among components are aligned, independent of their implementation platforms.

Admittedly, the level of interoperability of an MS&A tool with other tools and systems will be largely influenced by its compliance to standards. Interoperability is addressed separately to highlight the need in the homeland security application context. A high level of interoperability with other commonly used tools and systems is an indicator that the system is not duplicating any functions and it utilizes commonly available capabilities of other tools and systems. This also includes the capability of the tools to configure using data describing the scenario of interest. This improves the possibility of its use at other sites that use a similar set of tools and systems.

Available guidance

Improving interoperability of homeland security applications requires the development of the necessary infrastructure by the involved community. This includes common reference models, data dictionaries, glossaries, taxonomies, ontologies, and an architecture framework. The National Information Exchange Model (NIEM), an initiative under partnership between DHS, U.S. Department of Justice, and U.S. Department of Human and Health Services, "is designed to develop, disseminate and support enterprise-wide information exchange standards and processes that can enable jurisdictions to effectively share critical information in emergency situations, as well as support the day-to-day operations of agencies throughout the nation" (NIEM 2011). DHS has sponsored work in this area and is beginning to develop a coordinated approach. Jain and McLean (2008) define a component based architecture framework for simulation and gaming for incident management. McLean et al. (2008) provide a taxonomy for homeland security MS&A applications.

A few standards are available and efforts are in progress to develop more for the simulation interoperability area. The available standards include those for the High Level Architecture (HLA; IEEE 2000) and Distributed Interactive Simulation (DIS; IEEE 1998). DIS defines the formats of detailed level messages exchanged between distributed simulation while HLA is a software architecture with a defined Application Programmers Interface (API). There are efforts underway to update both DIS and HLA through SISO product development groups.

Another notable effort underway is focused on developing interoperability standards for commercial-of-the-shelf (COTS) simulation packages (CSPs). A product development group under the auspices of SISO is developing guidelines for integrating models implemented using various CSPs. A first standard addressing reference models for common CSP interoperability problems has been approved in 2010(Taylor et al. 2010, SISO 2010b). Four types of interoperability reference models have been defined addressing entity transfer, shared event, shared resources and shared data structures between simulations. An initial implementation utilizes grid computing and HLA functionality implemented using web services.

The National Infrastructure Simulation and Analysis Center (NISAC), a program under the United States Department of Homeland Security's Information Analysis and Infrastructure

Protection (IAIP) directorate, utilizes multiple integrated simulations of infrastructures for analyzing impact of man-made and natural incidents. NISAC utilizes Interoperable Distributed Simulation (IDSim) framework built using the reference implementation of the Open Grid Services Infrastructure (OGSI), Globus 3.2.1 (Linebarger et al. 2007). The framework has been used to integrate HLA federations with other simulations set up as IDSim federates.

MS&A applications can be considered as information systems. With that perspective, the Levels of Information Systems Interoperability (LISI) model can be considered relevant. The LISI model was developed by DoD Command, Control, Communications, Computer, Intelligence, Surveillance, and Reconnaissance group (C4ISR 1998). The model can be used to develop an interoperability profile for an information system. It depicts five levels of interoperability, isolated, connected, functional, domain, and enterprise, using four attributes, procedures, applications, infrastructure and data. The model may need to be updated to remove biases based on its development in 1998 (SEI 2009).

Recommended Implementation

Improved interoperability will considerably facilitate the integration of simulation models, the integration of M&S tool components, and the integration of M&S tools in the application environment in the M&S tool development and deployment process (shown in figures 2-3 and 2-4).

The available standards and developing standards should be used up to the extent possible to improve the chances of integrating multiple homeland security MS&A applications that may be needed to study and analyze actual or potential events and responses. Development teams should strive for achieving higher levels of the conceptual interoperability model presented in Turnitsa (2005). While the COTS simulation package interoperability standard effort is in progress, the available information on interoperability reference models should be used to guide the development. The evolved versions of HLA and DIS should be used for guidance wherever the implementation includes use of distributed simulations.

Use for Legacy vs New Applications

All new MS&A applications should be designed and developed to be interoperable with identified applications that they may be required to integrate with to fulfill their intended use. If other intended uses are envisaged in near term, these should be included when defining the interoperability requirements. Legacy applications would need to be modified to allow them to be interoperable with other applications. Software artifacts such as adapters or wrappers may need to be built around the legacy applications to allow them to interoperate with newer applications. Such an effort could be significant but may be required to allow meeting homeland security objectives.

Roles and responsibilities

Potential users and subject matter experts guiding the development effort for new applications or modification effort for legacy applications should identify other systems and applications that the

subject application will be required to interoperate with to fulfill its intended use. Developers need to identify the level of interoperability required with the other identified systems and applications, the applicable standards, and design, implement, and test the application or modification accordingly. Developers should ensure that the validation effort should include simulation executions in operational settings with the MS&A application interoperating with other operational systems called for by its intended use. Program managers with responsibility to acquire the MS&A application should ensure that interoperability requirements are specified, the plans call for identifying and implementing the relevant standards, and the plans are implemented.
They should also confirm that the test and evaluation of completed application includes tests for interoperability.

Costs/benefits

Interoperability of MS&A for homeland security applications can enable their reuse and in turn significantly reduce the development and implementation efforts. Interoperable applications can be implemented at multiple sites and thus reduce the development efforts that are executed currently for developing ad hoc applications. A large part of implementing MS&A applications goes towards integrating the application with a number of other systems including the operational interfaces that the target users employ in everyday operations and the systems that provide the input data required for MS&A. While a study with quantified estimates of the potential benefits of interoperability in MS&A application was not located, studies in other areas suggest that the benefits could be substantial. A study estimated that cost of inadequate interoperability of information technology systems in U.S. capital facilities industries at $15.8 billion per year (Gallaher et al. 2004). A NASA study in 2005 on cost benefit analysis of geospatial interoperability standards concluded that standards-based projects have a 119% ROI over the programs that did not implement standards (Longhorn and Blakemore 2007).

Metrics

Interoperability of MS&A applications may be assessed in two ways, standalone and pairwise. Standalone interoperability may be defined against a set of characteristics that facilitate an application being interoperable with other systems. These include use of neutral interfaces such as those using XML, and compliance with standards for distributed simulation such as HLA and DIS that may be applicable in their intended use environment. The pairwise interoperability may be defined among two MS&A applications that have to be integrated together for successful application in their common intended use environment. The level of conceptual interoperability model (LCIM) documented in Turnitsa (2005) may be used as a metric to assess the pairwise interoperability of MS&A applications.

The assessment of both standalone and pairwise interoperability requires identifying the MS&A applications, the systems or applications to be integrated, the operational systems, and the data sources that may be typically used in corresponding intended use environments. The intended use environment may be classified using the taxonomy for MS&A for homeland security applications defined by McLean et al (2008).

Practice Conclusion

Development of interoperable components allows efficient use of resources and is a recommended practice. The homeland security M&S community needs to work in a coordinated manner to develop the required infrastructure for achieving interoperability. Developers of M&S for homeland security applications need to employ practices that support interoperability such as use of XML interfaces, service-oriented architectures, and web services based on standards.

3.7 Execution Performance

Practice Introduction

The execution performance of an MS&A tool can be judged on multiple aspects including the execution time, the response times to various queries, and the hardware platform requirements. A system that executes quickly, responds to users' queries promptly, and does not require unique and expensive hardware to execute allows wider use is preferred. It has to be understood that trade-offs have to be made between execution time, cost and fidelity. Generally, quick turnaround times may be achieved at low cost but for a model with low fidelity. High fidelity models generally require longer execution times and/or expensive hardware platforms.

The performance requirements differ based on the application type, that is, for training, operations planning, trans-incident operational support etc. The application needs to simulate phenomenon in real time for training, and much faster than real time for trans-incident real time operations support purposes. At times, approximations or effects based simulation is used in training applications to meet the need to execute the simulation in real time. The execution time is generally not a constraint for operations planning applications, but an application executing much slower than real time may see limited use.

It should be noted that a combination of tools may be used to meet the time constraints. The National Atmospheric Release Advisory Center (NARAC) uses tools with short computation times to provide approximate predictions on plume dispersion within minutes following an incident and follows it with more accurate predictions based on outputs of tools that require longer computation times (LLNL, 2009). Such a strategy is used due to the longer computation times of the tools that provide accurate predictions. If the execution performance of the tools can be significantly improved, it will be possible to provide more accurate predictions to support response operations and thus improve the response capabilities.

The execution performance improvement for simulation models are usually achieved by utilizing high performance computing (HPC) platforms. It involves parallelizing the code and distributing it across multiple processors of a HPC platform. In previous years, there were two clearly different HPC architectures, one used multiple processors each with its own dedicated memory while the other used multiple identical processors with access to a common shared memory. The latter architecture is commonly referred to as symmetric multi-processor (SMP). Recently, platforms with hybrid architectures have become more common. These platforms may use a cluster of SMP machines, thus having nodes with their dedicated memory but within each node having multiple processors that are sharing the memory at that node. Methods have been developed for improving code execution speed utilizing the original two architectures and the more recent hybrid architectures. These methods can be utilized for improving execution performance of the simulation applications.

Available guidance

The task of utilizing HPC platforms for improved execution performance requires high expertise in the phenomenon being modeled, its representation in the code, and parallelization of codes. The procedures to parallelize a simulation code have to be usually specifically designed for the specific code and for the hardware configuration that the code is intended to run on. Given the need to design procedures specific to a code at hand, only a couple standards are available to guide the process.

There are de facto standards for two HPC platform architectures used for running parallelized codes on distributed processors. The Message Passing Interface (MPI) (Message Passing Interface Forum 2008) is currently the de facto standard for message-passing parallel programming implemented on distributed processors with dedicated memory spaces (Chorley et al. 2009). MPI offers a standard library interface that promotes the portability of parallel code whilst allowing hardware vendors to optimize the communication code to suit particular hardware.

More recently, compilers have been developed that are able parallelize some parts of the code for execution on processors with shared memory. The OpenMP (OpenMP Architecture Review Board 2007) is the current preferred standard for shared-memory programming. OpenMP offers a simple and yet powerful method of specifying work sharing between program threads, that leaves much of the low-level parallelization to the compiler.

Researchers are using hybrid MPI/OpenMP approaches for improving code execution performance on the hybrid hardware architectures that utilize SMP machines as nodes in a distributed platform. With the multiple evolving ways of setting up such hybrid architectures, no standard guidance has been developed.

Recommended Implementation

Improved execution performance will directly impact the production runs and analysis step in the simulation study process (shown in figure 2-2) and use of deployed tool step in the M&S tool development and deployment process (shown in figure 2-3). Execution performance improvement of the simulation code execution is a complex task and should be implemented for applications that can lead to clear and significant benefits. Clearly, the applications that are employed in trans-incident operational setting can lead to significant benefits from faster execution and the ability to provide results faster. Also, simulation models for other applications such as planning and training that take long time to execute, from hours to days, will gain from shorter execution times since that will make them more likely to be used.

The efforts should follow the current de facto standards, Message Passing Interface for distributed memory platforms and OpenMP for shared memory processors, and thus reduce efforts required for performance improvement. Both the standards have been employed for trans-incident operations applications. Key models at National Atmospheric Release Advisory Center (NARAC) have been parallelized using a combination of Message Passing Interface (MPI) and OpenMP, in order to support both multiprocessor and massively parallel computing platforms available to them (Nasstrom et al. 2007). Parallel processing has also been used at Los Alamos National Labs for running a year worth of simulation using EpiSims in less than 24 hours on a platform with 300 processors (Goetz 2006).

Use for Legacy vs New Applications

The practice to develop simulations with good execution time performance is recommended for all new applications. It is particularly recommended for the applications intended for time sensitive environments such as trans-incident operations support and training. The hardware costs continually go down allowing higher computing power within the same or lower budgets. Applications should hence be designed to exploit the new hardware architectures for the benefit of emergency response and for the wider homeland security needs.

The use of execution performance improvement techniques for legacy applications should be considered on a case by case basis. The effort to parallelize the code for legacy applications could be significant and in some cases may exceed the effort to develop the original legacy application. The execution performance improvement may be aligned with a major redevelopment effort, if the need for one is identified for other functional reasons.

Roles and responsibilities

The implementation of this practice will require access to highly qualified resources for developing the applications capable of exploiting the HPC platforms. The program managers should ensure that there is a clear need for rapid execution of the simulation application and associated expected benefits to justify the large budget required for harnessing the highly qualified resources and acquire HPC platforms or access and time to such platforms. The intended users would need to participate actively in identifying requirements that make fast execution of the applications a critical success criterion. They need to also support the program managers in developing the business case for use of such applications. The developers of applications with good execution performance would need to identify the right hardware architecture that can be accessible for the intended use sites and utilize programming methods that are appropriate for the selected hardware architecture.

Costs/benefits

A study on cost benefit of high performance computing estimated a benefit cost ratio in the range of 2.6 to 4.6 in research environment such as a national laboratory, and around 1.4 in industrial environment (Tichenor and Reuther, 2006). The benefit of a HPC system considered in the estimation for a research environment considered time saved by engineers and scientists in solving advanced problems, while the costs considered included the cost of the system, the time required to train users on it, prepare the application code(s) for parallel processing, launch the application(s), and administer the system. In an industry environment, the benefit considered was the estimated increased profit, while the costs included cost of the system, training costs, administration costs, and the cost of the software.

The above cost benefit ratios have been estimated for high performance computing in general and not specifically for improved execution performance of simulations. A number of researchers have identified the need for cost benefit analysis of parallel and distributed simulation to support their increased use across a wide range of environments.

Metrics

The metrics for execution performance improvements can be defined on two aspects. One would focus on the execution performance improvement achieved for the specific simulation model using code optimization and parallelization. The improvement is generally expressed as a ratio of old execution time to new execution time.

A second metric would be applicable in case of change in hardware for execution of the simulation model. A set of benchmarks, known as NAS Parallel Benchmarks (NPB), were developed for performance evaluation of parallel supercomputers by the Numerical Aerodynamic Simulation (NAS) program at NASA. These benchmarks suites can be executed on a hardware and the execution times can be used to evaluate its performance. Recent versions of the benchmarks are NPB 3.0 and GridNPB 3. NPB 3.0 is a set of implementation addressing OpenMP, High Performance Fortran (HPF), and Java, while GridNPB 3 is specifically designed to rate the performance of computational grids. The benchmarks are available on-line from NASA website (NASA 2009).

Practice Conclusion

The practice to improve execution performance of simulation models is recommended across all homeland security applications. It is highly recommended for simulation models that are designed for trans-incident operational support applications since faster results for such applications may lead to better response decisions that in turn may lead to reduced loss of lives and property. The execution performance improvements may be achieved through code optimization, distributing and parallelizing the code and through use of high performance computing (HPC) systems. Such improvements generally require high expertise and expensive equipment and hence a cost benefit analysis should be conducted before embarking on such an endeavor.

3.8 User friendliness and accessibility

Practice Introduction

The MS&A tools can be complex software applications and special attention needs to be placed on making them user friendly and ensuring that they have appropriate functionality and support for decision making. The practice of ensuring that products can be used by intended users to achieve their tasks is also referred to as human centered design.

The users for MS&A tools vary across the different application types that include analysis and decisions support, planning and operations, systems engineering and acquisition, and, training, exercises, and performance measurement (please see section 4 for more information on application types). Users in analysis and decisions support would generally have analytical background and be comfortable with MS&A applications. Users for systems engineering and acquisition applications are also generally anticipated to be comfortable with computer software and interfaces. Current and potential users of MS&A tools for planning and operations may have wide range of backgrounds including some with analyst experience who are comfortable with such tools but perhaps a larger group of people with emergency response and management backgrounds whose routine responsibilities involve limited interaction with computers. Similarly, current and potential users of MS&A tools for training, exercise, and performance measurement applications are expected to come from varying backgrounds with majority of them with limited experience with computer application and particularly with such tools.
Majority of targeted users for MS&A tools may have experience in emergency management and generally limited exposure to complex software applications per the above discussion. The goal of the MS&A tool developers should be to create interfaces that make them accessible to all users. MS&A tools should be embedded within or seamlessly interfaced with operational systems commonly employed by the targeted users in their daily routine. The users should have easy mechanisms to trigger the MS&A tools and should have to input minimum number of decision parameters. The tools should be set to access all the input data automatically from available sources. The use of earlier recommended practices of use of standards and interoperability will facilitate setting up automatic interfaces to data systems and to other models.

The outputs of MS&A tools can be complex and they have to be disseminated appropriately to the user that may include incident management organizations for planning and operation and training, exercises, and performance measurement applications. The preferred dissemination mechanism for complex MS&A tools for operations application may be a reachback center that provides expertise in deciphering the outputs for use by the incident management personnel. For critical decisions, procedures should be defined for accrediting the analysts to ensure that they have the expertise to correctly use the MS&A tools and interpret its results. Such analyst accreditation has been suggested earlier for tools modeling the nuclear stockpile (Trucano and Moya, 1999). In other cases, user friendly interfaces may be built to ensure rapid and correct understanding of the MS&A tool outputs by the incident management personnel. Also, training should be provided to the incident management personnel in the use of the tools and interpretation of their results.

Of course, user friendly interfaces should also be built for users with quantitative and analytical background who may be focusing on analysis and decision support and systems engineering and acquisition applications of MS&A tools. Such user would generally not require support of experts for deciphering the results.

Available guidance

The user friendliness of MS&A applications can benefit from guidance available for computer systems in general. A set of international standards, ISO9241: Ergonomics of human system interaction, defines a number of aspects for computer interfaces including those for displays, input devices, menu dialogs, command dialogs and software accessibility (ISO 2001). One of the included aspects is tactile and haptic interactions that may be relevant for first person gaming and simulation tools. Another standard, ISO 13407: Human-centered design for interactive systems, provides guidance on activities for implementing a system with human centered design (ISO 1999). A follow-up publication ISO TR18529: Ergonomics -- Ergonomics of human-system interaction -- Human-centered lifecycle process descriptions, provides a usability maturity model that describes seven processes each of which contains a set of base practices (ISO 2000).

In addition to the means for interactions with the application, the content of the outputs provided by MS&A applications will have a large impact on their usability. The outputs should quantify the uncertainty associated with the results. The users for analysis and decision support and systems engineering and acquisition applications are expected to appreciate the uncertainty information. The incident management personnel involved in planning and operations may look for a clear yes or no answer while the MS&A tools may provide answers with associated uncertainties and sensitivities. Ideally in an operations application, the decision makers should have had experience deciphering the outputs and the associated uncertainties based on several exercise cycles utilizing the specific MS&A tools. They should be able to incorporate the uncertainty and sensitivity information in their decision making. Going forward, the training, exercise, and performance measurement applications should incorporate the uncertainty information and thus prepare the users for planning and operation applications.

The uncertainty in MS&A tool results may be identified using uncertainty propagation techniques or using sensitivity analysis. The NASA standard on modeling and simulation (NASA 2008) calls for documentation of the process used for uncertainty quantification and the results in a simulation study. Pilch et al (2006) provide an overview of a methodology, quantification of margins and uncertainty (QMU), for assessing the uncertainty in simulation results. The ratio of margin and uncertainty is used to provide a measure of confidence that modeled system will work as expected.

Recommended Implementation

Improved user friendliness will significantly impact the production runs and analysis and the documentation and reporting steps in a simulation study (shown in figure 2-2). Similarly, it will largely impact the use of the deployed tool in the M&S tool development and deployment process (shown in figure 2-3).

The requirements for MS&A applications should consider the aspects of human computer interactions defined in ISO9241. Similarly, the development plans for the MS&A applications should comprehend the activities defined in ISO 13407 and ISO TR 18259.

Special attention should be paid to the users for planning and operations and training, exercises, and performance measurement applications. The results should be presented to the incident management organizations using visualizations and terminology familiar to them and should include information on uncertainty associated with the results. MS&A tools should use geographic map displays for relevant outputs or interface with post processing tools that allow such displays. The dissemination of outputs thus would gain from use of standards and a high level of integration with other tools and systems.

MS&A tools should comply with section 508 of the Rehabilitation Act of 1973 (USGSA 2009). The act requires that information systems should be accessible to Federal employees with disabilities. The user interfaces for MS&A tools for homeland security applications should be developed with this consideration.

Use for Legacy vs New Applications

The need to make applications easy to use exists for both new and legacy applications. The requirements for new applications should include those for user interfaces and interactions and take into account the background and skill levels of the intended users. User interfaces and outputs of legacy applications should be thoroughly reviewed to identify opportunities for improvements in response time and quality of decisions based on the interaction with and results of the applications. It is usually possible to develop wrap around software that provides an improved interface for interactions with the applications. Similarly, it is also usually possible to add post processing software that can take outputs of the MS&A applications and present it using map displays. The addition of uncertainty information in the results may be a bit more challenging but advisable for improved quality of decisions.

Roles and responsibilities

The implementation of this practice requires deep involvement of representative intended users at the requirements stage for both new applications and medication of legacy applications. If the recommended spiral approach is used, the users need to actively provide feedback on the friendliness of the applications and their outputs during each iteration of the prototype and the final product. Program and project managers should ensure that such involvement of users is in the project plan and that the planned activities are executed. The developers should actively solicit input on the user friendliness aspects of the applications at every planned interaction with the target users throughout the development cycle.

Costs/benefits

The cost of human centered design includes the cost of people on the development team who apply the methods for the purpose. The benefits of such methods can extend on multiple aspects including reduced development costs, improved functionality, reduced learning times, reduced task times, and reduced errors. Some case studies have been carried out to determine the costs and benefits of human centered design. Such studies compare the cost of implementing the methods for human centered design with the alternative of making the changes and fixes identified by the methods at a later stage. A study of implementation of a mission planning center at Israel Aircraft Industries indicated a cost benefit ratio of 1:15, and a study of joint application design process of the Inland Revenue in United Kingdom indicated a cost benefit ratio of 1:2.6 (Bevan 2005). While the ratios vary quite a bit, it is clear that the benefits can be significant.

Metrics

The system usability scale (SUS) may be used for a high level subjective view of usability and for comparison of usability among systems (Brooke 1996). It allows arriving at a score on a scale of 1-100 for the usability of a system. A more detailed assessment may be carried out that takes in to account the context of use of the system including the intended users, their objectives in using the system and the environment in which the system is being used. The maturity model defined in ISO TR 18529 should be used to assess the processes used for human centered design.

Practice Conclusion

MS&A applications should have appropriate functionality and interfaces based on intended use and users. They should provide their outputs such that they support the user in making the right decisions. There should be special attention paid to making the MS&A tools accessible to users with limited exposure to complex software applications such as those involved in operations and exercises. The applications for operational use should allow rapid arrival at the outputs to support decision making. The outputs themselves should be designed for quick understanding by the users. Careful consideration of the task and the intended users in the development of the MS&A applications can significantly reduce development cost and improve performance.

4 RELEVANCE OF THE BEST PRACTICES TO MS&A APPLICATION TYPES

In this section the best practices are discussed with reference to the four major application types including, analysis and decisions support, planning and operations, systems engineering and acquisition, and, training, exercises, and performance measurement. It is noted that the applications types are not mutually exclusive with respect to MS&A tools. The same MS&A tools can be used for multiple application types. For example, a model can be used for training applications to show the effect of decisions made by trainees, and it can also be used for operations application to select best plan of actions through evaluation of multiple alternatives using simulations. Additional best practices based on the application type are discussed where possible.

The relevance of the best practices may differ by MS&A application types. For example, it is a best practice to verify, validate and accredit MS&A tools. The verification, validation and accreditation (VV&A) procedures may be applied differently across the MS&A application types. VV&A of MS&A tools for application in training, exercises and performance measurement needs to focus on having the simulations contribute to development of an environment that appears as close to real life as possible. The VV&A of tools for application in planning and operations, that is, to support decision making in emergency response, needs to focus on the simulations' ability to predict the potential sequence of events. Ideally, MS&A tools for different applications should be verified and validated to represent the real world in all possible aspects. Unfortunately, the computing sciences and technology have not got to that point yet. For example, calculations to predict with high accuracy damages due to an explosion cannot be completed within the time it takes to have an explosion occur in a training scenario. Hence the option is to either pre-calculate the damages and allow the explosion to occur only at a predetermined place, or allow flexibility in the location of explosion and simulate damages that may be typical of such an explosion but not technically accurate. Pidd and Robinson (2007) discuss the focus on prediction versus insights for different simulation applications.

The applications and associated best practices are discussed based on the MS&A tools that were primarily designed and are used for that particular application type. Usually MS&A tools with training as the primary objective aim for running in real time and provide realism with a reduced emphasis on technical correctness. Alternatively, in training applications where technical correctness is critical, the calculations may be run offline ahead of time with the scenarios then restricted to the pre-calculated sequences. Tools for planning applications emphasize technical correctness and run time is not a major concern, while tools for operations need both technical correctness and execution time that are much faster than real time. Clearly, the tools designed for operations can be used for training and planning also. However, for the purpose of this report, the tools with both technical correctness and execution times that are much faster than real time will be considered as primarily for operations.

The relevance of the best practice to different application types is summarized in Table 3 (Table 1 presented in Executive Summary is reproduced as Table 3 for reader's convenience). Major aspects of the relevance assessments are discussed in the following sub-sections.

Table 3: Relevance of Best Practices to MS&A Application Types

Application Type ▶ Practice ▼	Analysis and Decision Support	Planning and Operations	Systems Engineering and Acquisition	Training, Exercises and Performance Measurement
Conceptual modeling practice	Similar emphasis across application types			
Innovative approaches	Similar emphasis across application types			
Software engineering practice	Similar emphasis across application types			
Model confidence/ Verification, validation and accreditation (VV&A)	Focus on technical correctness for M&S; Different for analysis tools	Critical; Focus on technical correctness	Focus on technical correctness	Focus on realistic appearance
Use of standards	Emphasized for integration	Critical for operation use; Emphasized for planning	Emphasized for integration	Need to include compliance to Sharable Content Object Reference Model (SCORM)
Interoperability	Emphasized for inputs and outputs	Critical for operation use; Emphasized for planning	Emphasized for inputs and outputs	Needed but not as much as for decision support or operations
Execution performance	Emphasized for allowing exploring multiple options	Critical for operations use; emphasized for planning	Needed at a level to support process	Critical to present realistic time responses
User friendliness and accessibility.	Emphasized to support decision making	Targeted to incident management personnel	Low need due to highly skilled users	Emphasized to support trainees

4.1 Analysis and decision support

As defined in section 2, analysis and decision support tools may be seen as focused on analysis rather than modeling and simulation (M&S). The best practices discussed in section 3 will need to be applied with different emphasis for the analysis and decision support application objective.

The emphasis will primarily be on dissemination of outputs since that is a key purpose for these tools. The other important practices are use of standards, interoperability, and performance. The VV&A practice may be used differently based on the type of the tool. A V&V methodology similar to that used for MS&A tools for earlier applications may be used for the cases where the tool is an environment with an embedded M&S tool, or for an M&S tool primarily used for reconstructing and analyzing incidents after they have occurred. A different V&V methodology may be used for analysis tools primarily to address errors in conceptual modeling, mathematical modeling, and where applicable, discretization, algorithm selection, programming activities, numerical solution, and representation of the output.

4.2 Planning and operations

All the best practices identified in section 3 for MS&A are applicable for the operations applications. The V&V aspect is critical for operations applications since MS&A tools outputs may support decisions that may affect human lives. The incident management organizations have to have a high degree of confidence in the outputs of the tools and that may be provided through the use of a rigorous V&V process. Along with the model validity, the need to support critical decisions also demands that the data used as input to the models should be the best available and should be validated to the extent possible.

The user friendliness of the MS&A tool, particularly of the interface available to the incident management personnel for understanding the tool's outputs is another critical aspect for operations applications of MS&A. The incident management decision makers should clearly understand the predictions provided by the model with the associated uncertainties before acting on them. Typically trans-incident operational support applications allow limited time to absorb the simulation outputs. Frequent exposure to MS&A tools output for operations planning applications will help the decision makers in their use of these tools for trans-incident operational support applications. Such exposure may be provided through exercises and integration of the tools where possible into day to day operational decision making.

A key aspect of use of MS&A for trans-incident operational support applications is the accessibility of such tools for incident management and response personnel. This requires additional infrastructure than mentioned earlier for providing the required validated data to MS&A tools. The required infrastructure may include centers of expertise that can serve as reachback for the incident management team and advise them on the predicted outputs regarding impact of the incident and evaluation of proposed response strategies. It may include platforms, tools, and expertise available right in the Emergency Operations Centers for such a role. It may also include hand held devices for the first responders for communication of results of MS&A tools such as the predicted path and affected areas for a toxic plume.

There are parallels in modeling incident management and military battles at tactical level. MS&A applications can gain from relevant advances in DoD simulation community. Daly and Tolk (2003) suggest use of applications traditionally found in the M&S community for defense command and control based on advanced information sharing and dissemination. Similar capabilities can be used to support command and control for emergency response efforts in reducing the risk of unintended consequences and efficient use of resources. M&S applications can be used to translate the volumes of data available into visual representations of current situation and potential scenarios. Such use of M&S applications requires high level of integration with other data systems, applications, and the command and control systems used for emergency response.

4.3 Systems engineering and acquisition

As defined in section 2, this application type includes use of MS&A tools for steps in the system engineering and acquisition processes. Among the best practices listed earlier there may be a relatively lower emphasis on performance since the systems engineering and acquisition applications would not be as time constrained as the operations support applications. Of course, these applications do have the constraint of completing the analysis in time to support the process. For example, the preliminary design needs to be verified in its ability to meet the functional requirement through the use of MS&A before the design is finalized and thus has limited time flexibility. Similarly there may be lower emphasis on the user friendliness and accessibility aspect since the primary users of the models for evaluation of product design may be subject matter experts familiar with the involved technology. The design evaluation models being custom developments may not need to emphasize compliance with interoperability requirements either. Most of the best practices would apply other than execution performance to the models used for evaluating the functioning of the system or equipment being acquired within the intended deployment environment. For example, the use of standards will facilitate integration of such tools as the Computer Aided Design (CAD) and Finite Element Modeling (FEM), and distributed simulations. Similarly, interoperability practice will facilitate the transfers of data across design and simulation tools used during the process.

4.4 Training, exercises, and performance measurement

Some of the best practices discussed in Section 3 may be employed a bit differently for assessing MS&A tools for homeland security training applications. The verification and validation of the tool may be to a different level than used for other application types. Ideally one would want to have the tools model the phenomena to the same level of detail and validate the outputs to the same level as for other application types, but the requirement for real time responses to the trainee may override this desire. For example, simulation of impact of a complex phenomenon like an explosion may require computation times much longer than the time it takes to occur in real life. The use of an explosion in a scenario for training hence has to be simulated with approximations to allow it to occur in realistic time durations or is pre-calculated off-line for use during the training. In first person serious games, "effects simulation" is used instead of simulation of the actual physics behind the phenomenon. Effects simulation consists of using approximate effects including visual impact (i.e., the explosion), and the resultant physical impact (i.e., the destruction and casualties). The approximations and effects simulations are used to allow as much reaction time to trainees as would be available to them in a real life incident and thus provide valuable training.

The interoperability best practice includes the capability of a tool to automatically configure the model to data read in from other systems. For a training application, the MS&A tools will be more valuable if the model can be customized to locations that trainees may actually be called to for an emergency situation. Some of the training tools provide fixed fictitious locations with several fixed parameters that provide limited value to the trainees due to the impression of a "toy problem" or an unrealistic "game." The interoperability level doesn't have to be as high as for tools supporting operations since the training application does allow the luxury of time to configure the model to an actual location and the scenario of interest. However, lower level of interoperability would translate to a higher level of effort required to customize the scenario and thus higher expense for training.

The standards practice for training applications should include compliance to Sharable Content Object Reference Model (SCORM). SCORM is a collection of standards and specifications for e-learning based training applications. The standards include specifications for communications between the host system, called the run-time environment, and the client side content. The SCORM specification was developed under the sponsorship of the United States Secretary of Defense as part of the Advanced Distributed Learning initiative. Bohl et al (2002) provide a critical assessment of SCORM.

The training applications may provide a good opportunity to familiarize the incident management personnel with the outputs generated by MS&A tools. While the first responder personnel may primarily train through interaction with simulation and gaming technologies to improve their response skills, the management personnel should be familiarized with MS&A tool outputs that they may be using in future for decision making. The training applications may also provide a good opportunity for collecting feedback on the user friendliness and accessibility of the tools used.

A number of considerations other than the list of the best practices presented in section 3 may be used to identify good training applications of MS&A. These may include:
- realism of the visualizations,
- value of the learning delivered by the tool,
- features for verification of learning,
- support for after action reviews, and,
- cost of the tool.

5 RELEVANCE OF THE BEST PRACTICES TO MS&A APPLICATION DOMAINS

In this section, the best practices are discussed in reference to major application domains. McLean, Jain, and Lee (2008) define following seven domains for MS&A for homeland security applications: social behavior, physical phenomena, environment, economic and financial, organizational, critical infrastructure, and other systems, equipment, and tools. Additional best practices based on the application domain are discussed where possible.

The relevance of the best practices may differ by MS&A application types. For example, it is a best practice to verify, validate, and accredit MS&A tools. The verification, validation, and accreditation (VV&A) procedures may be applied differently across the MS&A domains. VV&A of MS&A tools in social behavior domain focuses on plausibility and consistency. VV&A of physical phenomenon such as plume dispersion focuses on comparison of model results against measurements and hence in agreement with VV&A definition provided earlier.

The relevance of the best practice to different application types is summarized in table 4 (table 2 presented in Executive Summary is reproduced as table 4 for reader's convenience). Major aspects of the relevance assessments are discussed in the following sub-sections.

In general MS&A across all the application domains will gain from implementation of all the recommended best practices. The discussion below highlights the practices that are already being followed and those that are especially needed.

5.1 Social behavior

Modeling social behavior is a challenging endeavor. A common approach to model social behavior is through agent based models. Computational considerations generally lead to defining behavior rules for large groups of populations. Even if large computation times are allowed and a large set of behavior rules has been defined, it is difficult to predict outcomes with significant level of confidence.

It has been pointed out that the problem of validation is particularly difficult for models dealing with social sciences, particularly those that deal with group behavior (McNamara et al. 2008). The social behavior models are very difficult to validate due to the subject matter being human beings with their own individualities. Hutchings (2009) points out that for social science models the data is often subjective and the models are used to organize thinking and to study complex phenomena. He defines validation for such models as a check of consistency, plausibility, and the results making sense based on the understanding of the modeled system.

A recent report from the National Research Council (NRC 2008) identifies massively multiplayer online games (MMOGs) as an untapped resource for collecting social and behavioral data on a large scale. The report also recommends funding of research to determine the usefulness of MMOGs to verify and validate individual, organizational and societal models. This social behavior application domain will gain from such research.

Table 4: Relevance of Best Practices to MS&A Application Types

Application domain ▶ / Practice ▼	Social Behavior	Physical Phenomena	Environment	Economic and Financial	Organizational	Critical Infrastructure	Other Systems, Equipment, and Tools
Conceptual modeling practice	Important to select right paradigm			Need to emulate behavior and social processes		Important to select right paradigm	
Innovative approaches	Similar emphasis across application types						
Software engineering practice	Similar emphasis across application types						
Model confidence/ Verification, validation and accreditation (VV&A)	Focus on Plausibility	Comparison with measurements	Comparison with measurements	Focus on Plausibility	Focus on Plausibility	Comparison with measurements; good data sources available	Comparison with measurements
Use of standards	Need Standards	Need VV&A standards	Helped by standards	Need Standards	Need Standards; guides available	Need standards for cross sector models	Need VV&A Standards
Interoperability	Need efforts	Need efforts	Relatively better; need efforts	Need efforts		Critical to integrate models	Needed to integrate in other models
Execution performance	Needed due to use of ABM	High Performance Computing (HPC) platforms used		Needed due to use of Agent Based Modeling (ABM)		Needed for cross sector models	Not critical
User friendliness and accessibility	Needed to explain complex results	Helped by Geographical Information Systems (GIS) interfaces		Needed to explain complex results			Not critical

Social behavior models are often implemented using agent based modeling. The agent based modeling and simulation does not have a good set of defined standards. Grimm et al. (2006) highlight the lack of standards and propose a scheme for agent based modeling documentation

65

for promoting model transferability and reproducibility. As mentioned earlier, such models typically are constrained by computation times. This application domain will hence also gain from efforts aimed at implementing the practices for improving performance.

5.2 Physical phenomena

Physical phenomena models are somewhat more predictable than social behavior models but still quite difficult to validate. Hutchings (2009) distinguishes the physical science models as based on data collected by sensors and considered objective. He identifies the validation of such models as comparison of their results against measurements with consideration of uncertainty.

Physical phenomena models may achieve limited agreement with measurements given the high complexity of the modeled phenomena. In fact, one of the criteria to judge the validity of plume dispersion models is that their results agree with the tracer gas experiment results within a factor of 2, a low accuracy in the realm of discrete event simulation models for manufacturing. Researchers in the dispersion modeling area do not appear to agree on validation criteria and hence would gain from an effort for standard validation criteria and procedures.

A large number of MS&A tools exist for dispersion modeling contained in this domain. While some of the tools mentioned earlier appear to be interoperable to some extent as demonstrated by their successful integration with other applications, the majority appears to be comprised of standalone tools that lack interoperability. The user friendliness of these tools has been helped by availability of Geographical Information Systems (GIS) applications that allow graphical depiction of predicted dispersion over geographical areas.

Physical phenomena models are generally computation intensive and hence have attracted efforts to improve their execution performance using parallel and distributed simulations. The execution performance still continues to be a challenge for most MS&A tools in this domain leading to search for new ways.

5.3 Environment

Environment models are quite computation intensive, similar to physical phenomena models, and about as difficult to validate. There have been efforts to improve execution performance through use of high performance computers and parallel and distributed simulations. For example, the Weather Research and Forecast (WRF) model from the National Center for Atmospheric Research has been designed to execute on distributed, parallel and hybrid architecture computing platforms to achieve good execution performance (Michalakes et al. 2004). The WRF model also provide interoperability with community modeling infrastructure such as the Earth System Modeling Framework.

The user friendliness of outputs of these tools has gained from the wide interest across scientific and general population in weather forecasts. The outputs are typically available in GIS formats and increasingly incorporating 3D displays that allow understanding of various weather phenomena across the altitude. Xie et al. (2009) use a transformation component to export the output of their dust storm simulation model to GIS software through a web map service. The environment application domain has also gained from the availability of a number of standards for inputs and outputs, such as Binary Universal Form for Data Representation of Meteorological Data (WMO 2007) and Digital Weather Markup Language (NWS 2009).

5.4 Economic and financial

Economic and financial models share some of the characteristics of the social behavior models. A common criticism of economic models is that they assume rational behavior for all involved and that is not a good assumption. Hence similar to the social behavior models, they are hard to validate. Pidd (2006) suggests using Soft Systems Methodology (SSM) in conjunction with simulation for tackling problems involving entities with social behavior. SSM recommends use of social and political analysis to develop a good understanding of the problem situation that in turn allows development of simulation models that provide an improved representation.

The conceptual modeling practice, in particular the selection of the right modeling paradigm is important for the models in this domain for a good representation of the economic and financial behaviors. NISAC Agent Based Laboratory for Economics (N-ABLE) developed at Sandia National Laboratory utilizes agent based simulation methodology to model interdependencies between economic and infrastructure sectors (Eidson and Ehlen 2005). N-ABLE utilizes a cognitive-economic model of household behaviors that goes beyond traditional economic models to include psychological, non-market, and extreme-events effects that are important for homeland security applications (Ehlen et al. 2009).

5.5 Organizational

Organizational models are somewhat similar to social behavior models since they represent organizations comprised of humans. In fact, in addition to being subject to varying individual behavior, each organization has its own culture and perhaps multiple micro-cultures within its units, making them difficult to model and validate. Again, problem structuring methods such as SSM (Pidd 2006) will help in the development of such models. Rouse and Bodner (2009) point out that the utility of organizational simulations will largely depend on the degree to which they incorporate emulation of behavioral and social processes. Organizational simulation can be used for early identification of jurisdictional issues in context of homeland security applications.

Conceptual modeling practice is important for this domain since it covers social and behavioral aspects together with science and engineering. Successful organizational simulations will need to model the impact of leadership on the organizational behavior. The modeling task may be facilitated for organizations with well defined processes. The organizations involved in homeland security have to follow guides such as the National Response Framework and the National Incident Management System that help develop the models with some consistency. The underlying assumption is that the organizations follow the defined guides and procedures rationally. The assumption may hold to a somewhat larger extent than the rational behaviors of individual in economic models.

Cooper et al. (2005) report on their use of quantifiable mathematical analysis for modeling performance of organizations, teams, and personnel. The latent effects models developed by them may be used within simulation models of organization performance.

The organizational simulations would typically be agent based simulations and thus computationally intensive. These simulations will also need efforts to improve the execution performance to encourage increased usage.

5.6 Critical infrastructure

The critical infrastructure models also face complexity in validation due to the multiple interconnections among the infrastructures they represent. Multiple paradigms have been used to model critical infrastructures including system dynamics, continuous simulations, and discrete event simulations. The use of multiple paradigms contributes to complexity of validation.

Interoperability is important for the critical infrastructure models since these models will typically be used in an integrated manner. Multiple infrastructure models may be integrated together to see the cascading impact of disruption in one critical infrastructure to others. The infrastructure models may also be integrated with economic and social behavior models to study the impact of disruption in a critical infrastructure.

No standards are currently available that address infrastructure and cross sector modeling. The cross sector modeling issues are particularly complex due to multiple time scales and granularities that may be used in models of difference infrastructure (Pederson et al. 2006). This domain has benefited from efforts to develop and validate detail infrastructure and demographic data sources. Developers of models in this domain use the following two data sources: LandScan series of data sets maintained by Oak Ridge National Laboratory and National Asset Database set up by the Office of Infrastructure Protection (DHS/IP).

5.7 Other systems, equipment, and tools

Models for other systems, equipment, and tools share some of the concerns raised for models for evaluation of systems and equipment in the systems engineering and acquisition application type. These models will usually be specific to the system, equipment, or tool they represent. For example, Felsmann et al. (2009) describe validation methods and data sets for simulation of HVAC mechanical equipment. Such models may be useful in homeland security context for simulating entry and distribution of harmful agents through the HVAC system of a building.

The models of systems, equipment and tools do need to be designed for interoperability since they will generally be integrated into models of the environment they are deployed in. Department of Defense is at the leading edge in using real and simulated equipment simulators integrated through Distributed Interactive Simulation (DIS) framework for training. In the homeland security context, the first responders may use simulated equipment for training. The equipment simulations hence have to be developed to allow their integration into larger simulations.

6 APPLYING THE MS&A BEST PRACTICES

The intent of this section is not to identify the best projects or tools. Such identification would require a comparative assessment of all projects and tools in the selected domain and that is outside the scope of this effort. The projects and tools discussed in this section are used only to provide a context for the application of the best practices defined in the earlier sections.

6.1 TELL (Training, Exercise, and Lessons Learned) system

A good example of use of MS&A tools for homeland security training application is the DHS-sponsored TELL (Training, Exercise, and Lessons Learned) system developed with participation from Sandia National Labs, Lawrence Livermore National Labs, University of Southern California's Institute for Creative Technologies, and the Incident Management Consortium. The TELL system provides capabilities for executing an exercise, inserting injects generated by other simulation software, and capturing the proceedings for after action review. The system was used in September 2007 for a Golden Guardian preparatory exercise at the Anaheim Emergency Operations Center, Anaheim, CA. TELL was used to present a simulated catastrophic incident to an incident management team operating in a remote mobile incident command post. The ground truth and exercise injects were generated using Sandia's WMDDAC (Weapons of Mass Destruction Decision Analysis Center) and Lawrence Livermore's ACATS (Advanced Conflict and Tactical Simulation) software.

The simulation components of the TELL system have gone through verification and validation (V&V) efforts. ACATS is based on JCATS (Joint Conflict and Tactical Simulation) software that was developed for Department of Defense (DoD). While all the V&V details of JCATS do not appear to be publicly available, an available report for validation of non-lethal weapons (Taylor & Neta, 2001) suggests that V&V were carried out for JCATS.

The TELL system utilizes standards in some aspects and that places it in a good position with respect to integration with other tools. One of the primary objectives of TELL is to improve understanding of National Incident Management System (NIMS) and Incident Command System (ICS), both of which are official DHS standards. TELL utilizes standardized voice over Internet protocol (VOIP) for telephone and radio communications by trainees. The simulation software ACATS is capable of integrating with other simulations under the High Level Architecture (HLA), an IEEE standard. For the first exercise, the TELL system integrated multiple components including, WMDDAC, ACATS, Virtual News Network (VNN), Enterprise Virtual Operations Center (EVOC), event capture and record and the visualization modules.

The user friendliness and accessibility of outputs of the TELL system is helped by a team that includes personnel in two teams – simulation cell and inject cell. The outputs of simulations are analyzed and entered as injects by members of the inject cell team. The After Actions Review director and event recorder, both members of the simulation cell team, guide the trainees through understanding the lessons learned. The procedures associated with the use of the system thus ensure that the simulation outputs are correctly utilized in the training process.

The system has been used effectively for training and hence it has demonstrated that the included simulation models execute in required time frames for defined scenarios. No modifications to simulations were reported to meet the execution time constraints indicating a good performance by ACATS and WMDDAC, the involved software in this case.

The TELL team ensured that the trainees use exactly the same systems and interfaces that are available to them in real life. The capabilities of the simulation software provide the exercise controllers with a "God's eye" view. Sharing the "God's eye" view with trainees can give them a false expectation of the information available to them in a real life incident. The TELL team avoids this by having the trainees be at a mobile command post with their resident systems. This allows for the training to be totally transferable to real life application.

Overall, the TELL system supported by ACATS and WMDDAC simulation tools presents a good example of use of MS&A tools for training application. The visuals from the simulation software are presented to the trainees as video clips in the Virtual News Network newscasts and are quite realistic. The support for recording the exercise and the After Action Review allows for verification of learning by the trainees and ensures that lessons learned are captured. While no information is available on the cost of a TELL system based exercise, the training was reportedly found very useful by the trainees.

Another good example of use of interoperable MS&A tools for training is the Urban Chemical Disaster (UCD) federation developed by the Preparedness and Catastrophic Event Response (PACER) University Center of Excellence (UCE) sponsored by DHS. The UCD is an HLA federation that integrates multiple simulations including plume dispersion, traffic, sensors and command and control. The integrated capability has been used to train local decision makers on a chlorine rail car tank explosion in downtown Baltimore. PACER developed an M&S integration framework that can be used for integrating different simulations needed to represent an incident. The framework is also being used for integration of a Bioterrorism Crisis Management (BCM) simulation federation.

6.2 Pandemic influenza impact study

A recent important planning application of MS&A tools for homeland security purpose is the pandemic influenza impact study carried out by the National Infrastructure Simulation and Analysis Center (NISAC). NISAC utilized a number of MS&A tools to estimate the impact of pandemic influenza on national population, economy, and infrastructure (NISAC 2007). The study evaluated strategic alternatives to contain the impact of the pandemic influenza using MS&A tools and developed recommendations. The MS&A tools used included: two epidemiological simulation systems (EpiSimS and EpiCast) to determine the progression and overall impact of the pandemic on the population, translation of disease-modeling results into workforce absenteeism, national Critical Infrastructure Protection Decision Support System (CIPDSS) simulations, an aggregate national-scale system dynamics model of the U.S. electrical power system, a physics-based deterministic model of the electric power transmission grid, an operations and maintenance model of telecommunications network to simulate impact of workforce reduction, an estimation model for healthcare service impacts at the national and local levels, a hospital surge model for impacts on healthcare facilities, analysis of workforce absenteeism impacts in the transportation sector, analysis of impact on airfreight using air transportation optimization model (ATOM), analysis of delays in rail shipments using Rail Network Analysis System (R-NAS), a queuing model to analyze the impact of absenteeism on container ports, a system dynamics model of the milk supply chain to gain insights into impact on food supply chains, a firm level model developed using NISAC Agent-Based Laboratory for Economics (N-ABLE) for food value chain analysis, and macroeconomic simulations for analyzing economic impacts.

The multiple modeling paradigms utilized for the study allowed cross-model comparison and validation. While detailed information on validation of individual or integrated models was not located, it is fair to assume that such validations were performed based on the strong track record of the involved national labs for conducting validations of the models used in their nuclear weapons work.

Similarly, while there isn't explicit information available on the use of standards and level of interoperability of the tools, it is clear that a number of MS&A tools were used in an integrated manner to carry out the study. The use of multiple MS&A tools in an integrated manner allowed the study to develop recommendations across wide areas of interest including population, economy, and multiple infrastructures.

The NISAC study excels in the dissemination of the MS&A outputs. It includes an uncertainty analysis to help identify the affordable limits for planning for the pandemic influenza. It used 3 diverse epidemiological models to address the uncertainty in both disease characteristics and model structure. These three models were: EpiSimS, Like-Infect, and Generalized Infectious Disease (GID). The uncertainty analysis identified the range and distribution of the outcomes. A robustness evaluation using Loki-Infect compares and evaluates the effectiveness and costs of mitigation strategy combinations. A risk analysis that includes decision maker values is used to evaluate the trade-offs among alternative strategies. The simulation outputs are also used for a satisfaction and regret analysis to calculate the confidence level with which any intervention strategy can be selected over another. The study demonstrates the value of proper methods of dissemination of the output and thus of user friendliness and accessibility. It is not clear that the various analyses were built into the MS&A tools, but the operations planning application does allow the time to carry out such analysis for proper understanding by decision makers.

The execution performance metrics for the individual simulation components and integrated systems used in the NISAC study were not made available. However, execution performance is not critical for operations planning application. The report does indicate that hundreds of simulation runs were conducted suggesting that execution performance did not constrain the study.

Overall, the NISAC study on impacts of pandemic influenza presents a good example of use of MS&A tools for operations planning application. It is unique in the large scope it addressed in exploring the potential impact of the pandemic influenza. It clearly stands out in the multiple analyses of outputs of simulation runs for improved understanding of results and for developing insights.

6.3 IMAAC/NARAC

The Interagency Modeling and Atmospheric Assessment Center/ National Atmospheric Release Advisory Center (IMAAC/NARAC) at Lawrence Livermore National Lab provides a leading example of use of MS&A tools for trans-incident operational support application. The center is available to support incident management personnel responding to an emergency involving toxic plumes such as a fire at a chemical plant. IMAAC/ NARAC can model the toxic plume and provide initial results on its expected dispersion within minutes of receiving the call. It can provide updated results with increased accuracy within hours. The model outputs provide information on the areas that will be affected over time by the plume and the level of exposure of the population to the toxic agents over time. Such information is very useful to the incident management personnel in determining their response strategies including shelter-in-place and evacuation options.

NARAC staff use a number of integrated MS&A tools to come up with plume dispersion predictions. These include: a diagnostic meteorological model (ADAPT), a pressure effects model for high explosives and dirty bombs (BLAST), a mesoscale forecast model (COAMPS), a Gaussian plume model with hazardous chemical databases (EPICODE), a grid generation software for use by other models (GridGen), a Gaussian plume model for radioactive and nuclear material (HotSpot), a gross fission products fallout model (KDFOC), and a Lagrangian stochastic particle dispersion model (LODI).

NARAC stands out in ensuring that the models used are "extensively tested and evaluated," i.e., verified and validated to the extent possible. The simulations used are very complex and would require long execution times for high fidelity modeling. The response to emergencies does not allow the luxury of time for high fidelity modeling of the involved complex phenomenon. The tools used hence provide approximate solution with the degree of accuracy varying based on the computation time available. The tools hence are evaluated for appropriate degree of accuracy in their outputs. The MS&A tools used by NARAC have been evaluated using analytic solutions, field experiments, and operational testing. The analytic solutions compare the results to known exact results. The field experiments include comparison of model results with the actual path taken by tracer gas released in experimental locations. Operational testing involves comparison of model results with the measurements and records of actual past incidents.

The MS&A tools used at NARAC integrate with data sources and other tools using standard interfaces. The tools are integrated with systems providing current meteorological observations, weather forecasting systems, and geographical terrain elevation and population databases. NARAC is also collaborating for standardization and integration with other tools including with EPA/NOAA for CAMEO/ALOHA, and NRC for RASCAL. It is also integrating with outdoor-indoor infiltration models developed at Lawrence Berkeley National Lab (LBNL).

NARAC operates in a reachback mode for supporting incident management personnel. While the plume dispersion projections are disseminated through the web and client systems using graphical displays, analysts are available to explain the outputs to incident management personnel. This helps ensure that the outputs are correctly understood and used.

The MS&A tools at NARAC have been selected to meet execution times suitable to the purpose. The tools used for trans-incident operational support execute within a few minutes for initial results. NARAC does provide training and operations planning support and can use MS&A tools with longer execution time for such purposes.

Overall, NARAC tools are a good example of use of MS&A for homeland security trans-incident operational support application. They are unique as they are the only source authorized by DHS to provide plume dispersion predictions in case of an incident involving release of toxic agents. Such designation removes any confusion that the incident management personnel may have due to conflicting predictions from competing models.

The National Infrastructure Simulation and Analysis Center (NISAC) also provides trans-incident operational support by identifying potential impact of an incident on critical infrastructure. They provide prediction of impact of hurricanes to the incident management personnel in affected jurisdictions and operations centers (see NISAC 2010 for a list of pre-landfall analysis of hurricanes performed over the years). The analysis of impact by NISAC is based on a number of MS&A tools, some of which were mentioned above in section 3.1. Similar to NARAC, NISAC is a DHS authorized source for providing the support for infrastructure impact of an incident.

6.4 CIPDSS Decision Model

The *Critical Infrastructure Protection Decision Support System Decision Model (CIPDSS-DM) Tool* developed at ANL analyzes results of multiple simulation runs to identify the recommended alternatives based on the value structure and risk profile of the decision maker. Inputs to the tool include the results of simulation runs carried out for a various combination of parameters including the alternative response strategies, and the decision profile of the decision maker. The tool generates standard and parameterized decision maps and satisfaction-regret curves that can be used for understanding desirability of each alternative over others. It thus supports the selection of the appropriate alternative.

The tool excels in user friendliness and accessibility of the outputs to decision makers since the various methods used in the tool to present the desirability of different alternatives make for easy understanding of the simulation results. A user guide is available for explanation of the theory and use of the tool. The tool input files utilize the standard XML format. The underlying simulation, CIPDSS, has been the subject of a validation study conducted by LANL based on the impact of Hurricane Katrina on Baton Rouge, LA. The use of CIPDSS-DM on the pandemic influenza study would have subjected it to review by a large number of technical experts ensuring the analysis met the reasonableness test. It is also assumed that ANL would have subjected CIPDSS-DM to rigorous analysis for its use on the pandemic influenza study.

The CIPDSS-DM tool provides well defined interfaces and can be used for analyzing results from tools other than CIPDSS. It can be hence integrated easily to other tools. It can process simulation outputs rapidly to produce the decision maps and curves for decision support. Its use of the decision maps and satisfaction-regret curves is rather unique among reported analysis applications for simulation results.

Another good example of a good analysis tool based on M&S tool is the Learning Environment Simulator (LES), a decision support environment built on top of CIPDSS. The objective of LES is to bring simulation to decision makers to allow them to enhance their learning through frequent experimentation with different policy options in a range of scenarios. The mode of having analysts support the decision makers for analysis for simulation results limits the learning opportunities for the decision makers. The LES tool provides user friendly screens for input of data and output analysis related to a public health emergency.

7 CONCLUSION

This report recommended practices for use of MS&A tools for homeland security applications. The recommended practices include conceptual modeling practice, innovative approaches, software engineering practice, model confidence/ verification, validation and accreditation (VV&A), use of standards, interoperability, execution performance, and user friendliness and accessibility. The practices may be implemented with different emphasis based on the application type, namely, analysis and decisions support, planning and operations, system engineering and acquisition, and training, exercises and performance measurement. The report identified such differences in emphasis for different MS&A application types and for various MS&A application domains. It is hoped that this report will help in increasing the application of MS&A tools for homeland security applications and in improving the practices used for development and use of such tools.

Current and future proposed work includes an analysis of needs and requirements, a survey of tools, models, and datasets, and an analysis of available standards and gaps for MS&A tools for homeland security applications. The analysis of needs and requirements will help the enhancement of existing tools and development of new tools where needed. The survey would help potential users of MS&A tools in identifying suitable applications and availability of data sets to help evaluate them. Analysis of available standards will guide the developers in ensuring compliance with applicable standards to follow and thus apply the recommended practice. Analysis of gaps in availability of standards will help guide the M&S community interested in homeland security applications in identifying the critical needs and in mounting collaborative efforts for addressing them.

REFERENCES

1. AIAA, 1998. *Guide for Verification and Validation of Computational Fluid Dynamics Simulation*. Document# AIAA G-077. American Institute of Aeronautics and Astronautics, Reston, VA.

2. Ambrosiano, J., 2008. A Coarse-Grained Risk Methodology for Food Defense Derived from the CARVER + Shock Model. Presentation No. LA-UR-08-03514. Provided to NIST team on Aug. 27, 2008.

3. Anthony, S.D., M. Eyring, and L. Gibson, 2006. Mapping Your Innovation Strategy, *Harvard Business Review*, May2006, Vol. 84, Issue 5, 104-113.

4. Arthur, J. D., and R. E. Nance. 2007. Investigating the use of software requirements engineering techniques in simulation modelling. Journal of Simulation, 1: 159-174;doi:10.1057/palgrave.jos.425002.

5. ASME, 2006. *Guide for Verification and Validation in Computational Solid Mechanics*. Document# ASME V&V 10. New York, NY: ASME International.

6. Axelsson, J., 2002. Model Based Systems Engineering Using a Continuous-Time Extension of the Unified Modeling Language (UML). *Systems Engineering*. Vol. 5, Issue 3, 165 - 179.

7. Balci, O., 2003. Verification, Validation, and Certification of Modeling and Simulation Applications. In *Proceedings of the 2003 Winter Simulation Conference*, ed. S. Chick, P. J. Sánchez, D. Ferrin, and D. J. Morrice, 150-158. Piscataway, NJ: Institute of Electrical and Electronics Engineers.

8. Balci, O., 2004. Quality Assessment, Verification, and Validation of Modeling and Simulation Applications. In *Proceedings of the 2004 Winter Simulation Conference*, ed. R .G. Ingalls, M. D. Rossetti, J. S. Smith, and B. A. Peters, 122-129. Piscataway, NJ: Institute of Electrical and Electronics Engineers.

9. Balci, O., J.D. Arthur, and R.E. Nance, 2008. Accomplishing Reuse with a Simulation Conceptual Model. In *Proceedings of the 2008 Winter Simulation Conference*, ed. S. J. Mason, R. R. Hill, L. Mönch, O. Rose, T. Jefferson, and J. W. Fowler, 959 - 965. Piscataway, NJ: Institute of Electrical and Electronics Engineers.

10. Banks, J., J.S. Carson, B.L. Nelson, and D.M. Nicol, 2009. *Discrete-Event System Simulation*, Fifth edition. Upper Saddle River, NJ: Prentice Hall-Pearson Education Inc.

11. Blake, D., 2008. Data Validation. In *Workshop on Future Directions in Critical Infrastructure Modeling and Simulation, Final Report*, ed. N. Adam, October 28-30, 2008, Suffolk, VA.

12. Bohl, O., J. Scheuhase, R. Sengler, and U. Winand, 2005. The Sharable Content Object Reference Model (SCORM) - a Critical Review. In *Proceedings of the 2002 International Conference on Computers in Education*, Vol. 2, 950 – 951.

13. Bouyssou, D., T. Merchant, M. Pirlot, A. Tsoukias, P. Vincke, 2006. *Evaluation and Decision Models with Multiple Criteria: Stepping Stones for the Analyst*. New York, NY: Springer Science + Business Media Inc.

14. Brooke, J., 1996. SUS: a "Quick And Dirty" Usability Scale. In *Usability Evaluation in Industry*, ed. P. W. Jordan, B. Thomas, B. A. Weerdmeester and A. L. McClelland. London: Taylor and Francis.

15. Brown, M.J., M.D. Williams, G.E. Streit, M. Nelson, and S. Linger. 2007. An Operational Event Reconstruction Tool Used with Biological Agent Collectors for Source Inversion Applications. *Seventh Symposium on the Urban Environment*, 10-13 September 2007, San Diego, CA.

16. Brown, T., 2007. Multiple Modeling Approaches and Insights for Critical Infrastructure Protection. In *Computational Models of Risks to Infrastructure,* NATO Science for Peace and Security Series D: Information and Communication Security - Vol. 13. ed. D. Skanata and D.M. Byrd. 23-25. IOS Press 2007.

17. Chan, V., C. Musso, and V. Shankar, 2008. Assessing innovation metrics: McKinsey Global Survey Results. *McKinsey Quarterly*, Nov. 2008.

18. Chorley, M.J., D.W. Walker, and M.F. Guest, 2009. Hybrid Message-Passing and Shared-Memory Programming in a Molecular Dynamics Application on Multicore Clusters. *International Journal of High Performance Computing Applications*, Aug 2009, Vol. 23, 196 - 211.

19. Clemen, R.T., and T. Reilly, 2001. Making Hard Decisions with Decision Tools, Duxbury Press, Pacific Grove, CA.

20. Cooper, J.A., R. Robinett III, J. Covan, J. Brewer, L. Sena-Henderson, and R. Roginski, 2005. Final Report: Mathematical Method for Quantifying the Effectiveness of Management Strategies. Report SAND2004-6032. Sandia National Laboratories, Albuquerque, NM.

21. C4ISR, 1998. Levels of Information Systems Interoperability (LISI). C4ISR Interoperability Working Group, Department of Defense. Washington, D.C.

22. Daly, J.J., and A. Tolk, 2003. Modeling and Simulation Integration with Network-Centric Command and Control Architectures. Fall Simulation Interoperability Workshop 2003, Paper 03F-SIW-121, Orlando, Florida, September 2003. Available on-line via: http://www.vmasc.org/pubs/tolk-modeling01.pdf [last accessed April 13, 2009].

23. Dehoff, K., B. Jaruzelski, and A. Kandybin, 2007. Measuring Innovation. Public comment submission to U.S. Department of Commerce Advisory Committee on Measuring Innovation in the 21st Century Economy, May 11, 2007, Washington DC. Available on-line via: http://www.innovationmetrics.gov/comments/051107BoozAllenHamilton.pdf [last accessed on Oct. 1, 2009].

24. Deming, W.E. 1982. Out of the crisis. Cambridge: The MIT Press.

25. DoD, 1998. DoD Modeling and Simulation (M&S) Glossary. U.S. Department of Defense. Manual DoD 5000.59-M, January 1998. Available on-line via: http://www.dtic.mil/whs/directives/corres/pdf/500059m.pdf [last accessed Oct. 1, 2010].

26. DoD, 2006. VV&A Recommended Practices Guide (RPG). U.S. Department of Defense Modeling and Simulation Coordination Office. RPG Build 3.0. September 2006. Available on-line via: http://vva.msco.mil/ [last accessed Oct. 1, 2010].

27. DoD, 2007. DoD Modeling and Simulation (M&S) Management. U.S. Department of Defense. DoD Directive 5000.59. August 8, 2007. Available on-line via: http://www.dtic.mil/whs/directives/corres/pdf/500059p.pdf [last accessed Oct. 1, 2010].

28. DoD, 2009. DoD Modeling and Simulation (M&S) Verification, Validation, and Accreditation (VV&A). Department of Defense Instruction Number 5000.61. December 9, 2009. Available on-line via: http://www.dtic.mil/whs/directives/corres/pdf/500061p.pdf [last accessed October 4, 2010]

29. DON, 2004. Modeling and Simulation Verification, Validation, and Accreditation Implementation Handbook, Volume I: VV&A Framework, 30 March 2004. Prepared by: U.S. Navy Modeling and Simulation Management Office.

30. Dongarra, J., and F. Sullivan, 2000. Guest Editors' Introduction: The Top 10 Algorithms. Computing in Science and Engineering, Vol. 2, No. 1, p. 22-23.

31. Drake, J., I. Foster, B. Malone, D. Williams, and D. Bader, 2009. Global Systems Simulation Software Requirements: An Outline for CSET-ACPI Leveraging. Available on-line via: http://www.csm.ornl.gov/ACPI/Documents/ACPI_CSET.htm [last accessed June 30, 2009].

32. Ehlen, M.A., M.L. Bernard and A.J. Scholand, 2009. Cognitive Modeling of Household Economic Behaviors during Extreme Events. In: Social Computing and Behavioral Modeling. Eds: H. Liu, J. Salerno, and M.J. Young. Springer, New York.

33. Eidson, E.D. & M.A. Ehlen, 2005. NISAC Agent-Based Laboratory for Economics (N-ABLE™): Overview of Agent and Simulation Architectures. Sandia National Laboratories Technical Report SAND2005-0263, Albuquerque, New Mexico.

34. Feather, M.S., 2004. Towards Cost-Effective Reliability through Visualization of the Reliability Option Space. Annual reliability and maintainability symposium, Los Angeles CA, 26-29 January 2004, 546-552. Felsmann, C., J. Lebrun, V. Lemort, and A. Wijsman, 2009. Testing and Validation of Simulation Tools of HVAC Mechanical Equipment Including their Control Strategies. Part I: description of the validation test cases. Eleventh International IBPSA Conference, Glasgow, Scotland, July 27-30, 2009. 2059-2063.

35. Galin, D., and M. Avrahami, 2006. Are CMM Program Investments Beneficial? Analyzing Past Studies. IEEE Software, Vol. 23, Issue 6, Nov.-Dec. 2006. 81 - 87.

36. Gallaher, M.P., A.C. O'Connor, and T. Phelps, 2002. Economic Impact Assessment of the International Standard for the Exchange of Product Model Data (STEP) in Transportation Equipment Industries, National Institute of Standards and Technology (NIST) Planning Report 02-5.

37. Gallaher, M.P., A.C. O'Connor, J.L. Dettbarn, Jr. and L.T. Gilday, 2004. Cost Analysis of Inadequate Interoperability in the U.S. Capital Facilities Industry. National Institute of Standards and Technology. Report NIST GCR 04-867. Available online via: http://www.bfrl.nist.gov/oae/publications/gcrs/04867.pdf [last accessed on Aug. 4, 2009]

38. Gisler, G., R. Weaver, M.L. Gittings, 2006. Two-Dimensional Simulations of Explosive Eruptions of Kick-Em Jenny and Other Submarine Volcanoes. *Science of Tsunami Hazards*, Vol. 25, No. 1, p. 35-41.

39. Glickman, T.S., 2008. Program Portfolio Selection for Reducing Prioritized Security Risks. *European Journal of Operational Research*, Vol. 190, No. 1, 268-276.

40. Goetz, T., 2006. The Battle to Stop Bird Flu. *WIRED magazine*. Issue 14.01. January 2006. Available on-line via: http://www.wired.com/wired/archive/14.01/birdflu.html?pg=1&topic=birdflu&topic_set [last accessed Sept. 02, 2009]

41. Gordon, S.C., 2000. Determining the Value of Simulation. *Proceedings of the Summer Computer Simulation Conference*. ed. W. Waite, Society for Modeling and Computer Simulation International, 969-973.

42. Grimm, V., U. Berger, F. Bastiansen, S. Eliassen, V. Ginot, J. Giske, J. Goss-Custard, T. Grand, S. K. Heinz and G. Huse, 2006. A Standard Protocol for Describing Individual-Based and Agent-Based Models. *Ecological Modelling*. Vol. 198, No. 1-2. (15 September 2006), 115-126.

43. Guru, A. and P. Savory, 2004. A Template-Based Conceptual Modelling Infrastructure for Simulation of Physical Security Systems. In *Proceedings of the 2004 Winter Simulation Conference*, ed. R.G. Ingalls, M.D. Rossetti, J.S. Smith and B.A. Peters. 866-873. Piscataway, NJ: Institute of Electrical and Electronics Engineers.

44. Huang, E., K. Kwon, and L. McGinnis, 2008. Toward On-Demand Wafer Fab Simulation Using Formal Structure & Behavior Models. In *Proceedings of the 2008 Winter Simulation Conference*, ed. S. J. Mason, R. R. Hill, L. Mönch, O. Rose, T. Jefferson, and J. W. Fowler, 2341-2349. Piscataway, NJ: Institute of Electrical and Electronics Engineers.

45. Huizenga, D., 2007. Testimony on the "Radiation Detection" before the House Energy and Commerce Subcommittee on Oversight and Investigations. Available on line via http://www.nnsa.energy.gov/news/print/1228.htm [Last accessed on January 22, 2009]

46. Hutchings, C.W., 2009. Enabling Homeland Security with Modeling & Simulation (M&S). *Interservice/Industry Training, Simulation, and Education Conference (I/ITSEC)*, Paper ID# 9412, Orlando, FL, Nov. 30-Dec. 3.

47. IEEE, 1990. *IEEE Standard Glossary of Software Engineering Terminology. IEEE Std 610.12-1990.* Piscataway, NJ: Institute of Electrical and Electronics Engineers.

48. IEEE, 1996. *IEEE/EIA 12207.0-1996. Industry implementation of International Standard ISO/IEC 12207:1995.* Piscataway, NJ: Institute of Electrical and Electronics Engineers.

49. IEEE, 1997. *IEEE/EIA 12207.1-1997. Guide to IEEE/EIA 12207 - Software Life Cycle Processes - Life Cycle Data–and Implementation Considerations.* Piscataway, NJ: Institute of Electrical and Electronics Engineers.

50. IEEE, 1998. *IEEE 1278.1a-1998 Standard for distributed interactive simulation - application protocols.* Piscataway, NJ: Institute of Electrical and Electronics Engineers.

51. IEEE, 2000. *IEEE 1516-2000 standard for modeling and simulation (M&S) high level architecture (HLA) - framework and rules.* Piscataway, NJ: Institute of Electrical and Electronics Engineers.

52. IEEE, 2004. *Guide to the Software Engineering Body of Knowledge - 2004 Version.* Executive editors, Alain Abran, James W. Moore ; editors, Pierre Bourque, Robert Dupuis. IEEE Computer Society. ISBN 0-7695-2330-7. Available on-line via http://www.swebok.org [last accessed Oct. 1, 2010].

53. IEEE, 2007. *IEEE 1516.4-2007 Recommended Practice for Verification, Validation, and Accreditation of a Federation - an Overlay to the High Level Architecture Federation Development and Execution Process.* Piscataway, NJ: Institute of Electrical and Electronics Engineers.

54. INCOSE, 2008. Survey of Model-Based Systems Engineering (MBSE) Methodologies. Prepared by: ModelBased Systems Engineering (MBSE) Initiative, International Council on Systems Engineering (INCOSE). Document No.: INCOSE-TD-2007-003-01, Version/Revision: B, 10 June 2008. Available via: http://www.incose.org/ProductsPubs/pdf/techdata/MTTC/MBSE_Methodology_Survey_2008-0610_RevB-JAE2.pdf [accessed Jan. 13, 2011].

55. ISO, 1999. *ISO 13407: Human-Centered Design for Interactive Systems.* Geneva, Switzerland: International Organization for Standardization.

56. ISO, 2000. *ISO TR18529: Ergonomics -- Ergonomics of human-system interaction -- Human-centred lifecycle process descriptions.* Geneva, Switzerland: International Organization for Standardization.

57. ISO, 2001. *ISO 9241-1:1997/Amd 1:2001 – Ergonomics of Human System Interaction, Part 1: General Introduction.* Geneva, Switzerland: International Organization for Standardization.

58. Jain, S., and C.R. McLean, 2003. *Modeling and Simulation for Emergency Response: Workshop Report, Standards and Tools.* National Institute of Standards and Technology Interagency Report, NISTIR-7071.

59. Jain, S., and C.R. McLean, 2008. Components for an Incident Management Simulation and Gaming Framework and Related Developments. *SIMULATION*, Vol. 84, No. 1, Jan. 2008, 3-25.

60. Jain, S., C.R. McLean, Y. Tina Lee, 2007. Towards Standards for Integrated Gaming and Simulation for Incident Management. In *Proceedings of the 2007 Summer Computer Simulation Conference*. July 15-18, San Diego, CA.

61. Kuhl, F., R. Weatherly and J. Dahmann, 1999. *Creating Computer Simulations: An Introduction to the High Level Architecture*. Upper Saddle River, NJ: Prentice Hall.

62. Kussmaul, C. 2005. Using agile development methods to improve student writing. *Journal of Computing Sciences in Colleges*, Volume 20 Issue 3, 148-156.

63. Lacy, L.W., W. Randolph, B. Harris, S. Youngblood, J. Sheehan, R. Might and M. Metz, 2001. Developing a Consensus Perspective on Conceptual Models for Simulation Systems. In *Proceedings of the 2001 Spring Simulation Interoperability Workshop*.

64. LeClaire, R., G.B. Hirsch, and A. Bandlow, 2008. Learning Environment Simulator for First Responders. Presented at the LANL Risk Symposium 2008. March 13, 2008. Ref. No. LA-UR-08-1471.

65. Lee, E. K., S. Maheshwary, J. Mason, W. Glisson, 2006. Decision Support System for Mass Dispensing of Medications for Infectious Disease Outbreaks and Bioterrorist Attacks. *Annals of Operations Research*, Vol. 148, 25-53.

66. Lee, S.H., and W. Chen, 2009. A Comparative Study of Uncertainty Propagation Methods for Black-Box-Type Problems. *Structural and Multidisciplinary Optimization*. Vol. 37, 239–253.

67. Linebarger, J.M., M.E. Goldsby, D. Fellig, M.F. Hawley, P.D. Moore, and T.J. Sa, 2007. Smallpox over San Diego: Joint Real-Time Federations of Distributed Simulations and Simulation Users under a Common Scenario. In *Proceedings of 21st International Workshop on Principles of Advanced and Distributed Simulation (PADS'07)*. Available on line via: http://www.sandia.gov/nisac/docs/Smallpox%20over%20San%20Diego.pdf [last accessed on July 16, 2009].

68. LLNL, 2009. Emergency Response System: Real Time Operational Models. National Atmospheric Release Advisory Center, Lawrence Livermore National Laboratory, Livermore, CA. Available on-line via: https://narac.llnl.gov/modeling.php [last accessed on April 19, 2009].

69. Longhorn, R.A., and M. J. Blakemore, 2007. *Geographic Information: Value, Pricing, Production, and Consumption*. CRC Press. p. 198.

70. Mackerrow, E., 2003. Understanding Why – Dissecting Radical Islamist Terrorism with Agent-Based Simulation. *Los Alamos Science*, Number 28. Available on-line via: http://jtac.uchicago.edu/conferences/05/resources/ThreatAnticipationModel-MacKerrow-LosAlamos.pdf [last accessed on Jan. 2, 2009].

71. McLean, C.R., S. Jain, and Y.T. Lee, 2008. A Taxonomy of Homeland Security Modeling, Simulation, and Analysis Applications. *Spring Simulation Interoperability Workshop (SIW)*, Paper No. 08S-SIW-098. April 14-18, 2008, Providence, RI.

72. McLean, C.R., S. Jain, and Y.T. Lee, 2009. Overview of MSA Needs for Homeland Security. Interservice/ Industry Training, Simulation, and Education Conference (I/ITSEC) 2009. Paper No. 9505. Orlando, Nov. 30-Dec. 2, 2009.

73. McNamara, L.A., T.G. Trucano, G.A. Backus, S.A. Mitchell, and A. Slepoy, 2008. R&D for Computation Cognitive and Social Models: Foundations for Model Evaluation through Verification and

Validation (Final LDRD Report). Report SAND2008-6453. Sandia National Laboratories, Albuquerque, NM.

74. Message Passing Interface Forum, 2008. *MPI: A Message Passing Interface Standard Version 2.1*, available at http://www.mpi-forum.org/

75. Michalakes, J., J. Dudhia, D. Gill, T. Henderson, J. Klemp, W. Skamarock, and W. Wang, 2004. The Weather Research and Forecast Model: Software Architecture and Performance. In *Proceeding of the Eleventh ECMWF Workshop on the Use of High Performance Computing in Meteorology*, ed. George Mozdzynski. 25–29 October 2004, Reading, U.K.

76. Miller, I., R. Kossik, and C. Voss, 2003. General Requirements for Simulation Models in Waste Management. Waste Management 2003 Symposium, February 23-27, Tucson AZ, Available on-line via: http://citeseerx.ist.psu.edu/viewdoc/download?doi=10.1.1.123.8163&rep=rep1&type=pdf [last accessed June 30, 2009].

77. Nance, R.E. and J.D. Arthur, 2006. Software Requirements Engineering: Exploring the Role in Simulation Model Development. *Proceedings of the 2006 Operational Research Society Simulation Workshop (SW'06)*, March 2006, Coventry, England, 117-127.

78. NASA, 2008. *Standard for Models and Simulations.* NASA Technical Standard, NASA-STD-7009. Approved July 11, 2008. National Aeronautics and Space Administration, Washington, DC 20546-0001.

79. NASA, 2009. *NAS Parallel Benchmarks.* Available on-line via: http://www.nas.nasa.gov/Resources/Software/npb.html [accessed Sept. 10, 2009].

80. Nasstrom, J.S., G. Sugiyama, R. Baskett, S. Larsen, and M. Bradley, 2007. The National Atmospheric Release Advisory Center (NARAC) Modeling and Decision Support System for Radiological and Nuclear Emergency Preparedness and Response. NARAC Ref # UCRL-JRNL-211678-Rev2. *International Journal of Emergency Management (IJEM)*, Vol. 4, No. 3, 2007, 524-550.

81. NIEM, 2011. National Information Exchange Model. U.S. Department of Justice and U.S. Department of Homeland Security. Available via: http://www.niem.gov/ [accessed Jan. 21, 2011].

82. NISAC, 2007. National Population, Economic, and Infrastructure Impacts of Pandemic Influenza with Strategic Recommendations. National Infrastructure Simulation & Analysis Center (NISAC), Infrastructure Analysis and Strategy Division, Office of Infrastructure Protection, US Department of Homeland Security. October 2007.

83. NISAC, 2010. Fast Simulation and Analysis Team (FAST) Reports. National Infrastructure Simulation and Analysis Center. Available on-line via: http://www.sandia.gov/nisac/pub_fast.html [accessed February 23, 2010].

84. NRC, 1997. *Technology for the United States Navy and Marine Corps, 2000-2035*. Naval Studies Board. Commission on Physical Sciences, Mathematics, and Applications. Vol. 9: Modeling and Simulation, Chapter 2. National Research Council, National Academies Press, Washington, DC. Available on-line via: http://www.nap.edu/catalog.php?record_id=5869 [last accessed on Feb. 6, 2010].

85. NRC, 2006. *Defense Modeling, Simulation, and Analysis: Meeting the Challenge*. Authors: Committee on Modeling and Simulation for Defense Transformation, Board on Mathematical Sciences and Their Applications, National Research Council, National Academies Press, Washington, DC. Available on-line via: http://www.nap.edu/catalog.php?record_id=11726 [last accessed on Nov. 16, 2009].

86. NRC, 2008. *Behavioral Modeling and Simulation: From Individuals to Societies*. Authors: Greg L. Zacharias, Jean MacMillan, and Susan B. Van Hemel, Editors, Committee on Organizational Modeling: From Individuals to Societies, National Research Council, National Academies Press, Washington, DC.

Available on-line via; http://www.nap.edu/catalog.php?record_id=12169 [last accessed on Sept. 29, 2010]

87. NSF 2006. Simulation-Based Engineering Science: Revolutionizing Engineering Science through Simulation. Report of the National Science Foundation Blue Ribbon Panel on Simulation-Based Engineering Science. Available via: http://www.nsf.gov/pubs/reports/sbes_final_report.pdf [accessed Feb. 20, 2011].

88. Oberkampf, W.L., M. Pilch, and T.G. Trucano, 2007. Predictive Capability Maturity Model for Computational Modeling and Simulation. Sandia National Laboratories, SAND2007-5948.

89. OMG, 2009a. *Documents associated with UML Version 2.2.* Object Management Group. Available on-line via: http://www.omg.org/spec/UML/2.2/ [last accessed on April 19, 2009].

90. OMG, 2009b. *OMG Systems Modeling Language Version 1.1.* Object Management Group. Available on-line via: http://www.sysmlforum.com/docs/specs/OMGSysML-v1.1-08-11-01.pdf [last accessed on April 19, 2009].

91. OpenMP Architecture Review Board, 2007. *OpenMP Application Programming Interface, Version 2.5,* available at http://www.openmp.org.

92. Patenaude, A., 1996. Study on the Effectiveness of Modeling and Simulation in Weapons Systems Acquisition Process. Study conducted for the Deputy Director, Test, Systems Engineering and Evaluation, Office of the Secretary of Defense. Science Applications International Corporation, McLean, VA.

93. Paul, C., H.J. Thie, E. Reardon, D.W. Prine, and L. Smallman, 2006. Implementing and evaluating an innovative approach to simulation training acquisitions. RAND National Defense Research Institute, RAND Corporation, Santa Monica, CA.

94. Pederson, P., D. Dudenhoeffer. S. Hartley and M. Permann, 2006. Critical Infrastructure Interdependency Modeling: A Survey of U.S. and International Research. Idaho National Laboratory technical report INL/EXT-06-11464.

95. Pidd, M., 1999. Just Modeling Through: A Rough Guide to Modeling. *Interfaces*, Vol. 29, No. 2, 118-132.

96. Pidd, M., 2006. Making sure you tackle the right problem: linking hard and soft methods in simulation practice. In *Proceedings of the 2006 Winter Simulation Conference*, ed., S. G. Henderson, B. Biller, M.-H. Hsieh, J. Shortle, J. D. Tew, and R. R. Barton, 195-205. Piscataway, New Jersey: Institute of Electrical and Electronics Engineers.

97. Pidd, M., and S. Robinson. 2007. Organising Insights into Simulation Practice. In *Proceedings of the 2007 Winter Simulation Conference*, ed., S. G. Henderson, B. Biller, M.-H. Hsieh, J. Shortle, J. D. Tew, and R. R. Barton, 771-775. Piscataway, New Jersey: Institute of Electrical and Electronics Engineers.

98. Pilch, M., T.G. Trucano, and J.C. Helton, 2006. Ideas Underlying Quantification of Margins and Uncertainties (QMU): A White Paper. Report SAND2006-5001, Sandia National Laboratories, Albuquerque, NM.

99. Powell, J., 2006. Toward a Standard Benefit-Cost Methodology for Publicly Funded Science and Technology Programs. National Institute of Standards and Technology Interagency Report, NISTIR 7319.

100. Rao, M., S. Ramakrishnan, and C. Dagli, 2008. Modeling and simulation of net centric system of systems using systems modeling language and colored Petri-nets: A demonstration using the global earth observation system of systems. *Systems Engineering*, Vol. 11, No. 3, 203-220.

101. Rico, D. F., 2002. Software process improvement: Modeling return on investment (ROI). *2002 Software Engineering Institute (SEI) Software Engineering Process Group Conference* (SEPG 2002), Phoenix, Arizona.

102. Robinson, S., 2006. Issues in conceptual modelling for simulation: setting a research agenda. In *Proceedings of the 2006 Operational Research Society Simulation Workshop*, Eds: S. Robinson, S. Taylor, S. Brailsford, and J. Garnett. 165-174. Birmingham, England: Operational Research Society.

103. Rouse, W.B. and D.A. Bodner, 2009. Organizational Simulation. In: *Handbook of Systems Engineering and Management*, Eds: A.P. Sage and W.B. Rouse. Wiley, New York. 763-792.

104. Roy, C.J., and W.L. Oberkampf. 2010. A Complete Framework for Verification, Validation, and Uncertainty Quantification in Scientific Computing. 48th AIAA Aerospace Sciences Meeting Including the New Horizons Forum and Aerospace Exposition. Paper No. AIAA 2010-124. January 4 – 7, 2010, Orlando, Florida. Available via: http://www.aoe.vt.edu/~cjroy/Conference-Papers/AIAA-2010-124.pdf [accessed Jan. 19, 2011].

105. Royce, W., 2002. CMM vs. CMMI: From Conventional to Modern Software Management. *The Rational Edge*, February. Available on-line via: http://www.ibm.com/developerworks/rational/library/content/RationalEdge/feb02/ConventionalToModernFeb02.pdf[last accessed on March 30, 2009].

106. Rushby, J., 2006. Harnessing Disruptive Innovation in Formal Verification. *Proceedings of the Fourth IEEE International Conference on Software Engineering and Formal Methods (SEFM'06)*, IEEE Computer Society.

107. Samsa, M.E., J.C. VanKuiken, and M.J. Jusko, 2008. Critical Infrastructure Protection Decision Support System Decision Model: Overview and Quick-Start User's Guide. Argonne National Lab, Report ANL/DIS-08/7. September 2008.

108. Sargent, R.G., 2010. Verification and Validation of Simulation Models. In: *Proceedings of 2010 Winter Simulation Conference Proceedings*. Eds: B. Johansson, S. Jain, J. Montoya-Torres, J. Hugan, and E. Yücesan. 166-183. Piscataway, NJ: Institute of Electrical and Electronics Engineers.

109. Sargent, R.G., R.E. Nance, C.M. Overstreet, S. Robinson, and J. Talbot, 2006. The Simulation Project Life-Cycle: Models and Realities. In *Proceedings of the 2006 Winter Simulation Conference*, Eds: L. F. Perrone, F. P. Wieland, J. Liu, B. G. Lawson, D. M. Nicol, and R. M. Fujimoto. 863-871. Piscataway, NJ: Institute of Electrical and Electronics Engineers.

110. Schuyler, J., 2001. *Risk and Decision Analysis in Projects*, 2nd edition. Project Management Institute, Philadelphia, PA.

111. SEI, 2009. *Levels of Information Systems Interoperability*. Software Engineering Institute. Available on-line via: http://www.sei.cmu.edu/isis/guide/introduction/lisi.htm [last accessed on Aug. 4, 2009].

112. SISO, 1999. *Real-time Platform Reference Federation Object Model, SISO-STD-001.1-1999*. Simulation Interoperability Standards Organization, Inc., P.O. Box 781238, Orlando, FL 32878-1238, USA.

113. SISO, 2006. *Base Object Model (BOM) Template Specification, SISO-STD-003-2006*. Simulation Interoperability Standards Organization, Inc., P.O. Box 781238, Orlando, FL 32878-1238, USA.

114. SISO, 2010a. *GM-VV PDG - Generic Methodology for VV&A in the M&S Domain.* Simulation Interoperability Standards Organization, Inc., P.O. Box 781238, Orlando, FL 32878-1238, USA. Available via: http://www.sisostds.org/StandardsActivities/DevelopmentGroups/GMVVPDGGenericMethodologyforVVAintheM.aspx [last accessed: Jan. 19, 2011].

115. SISO, 2010b. *Standard for Commercial-off-the-shelf Simulation Package Interoperability Reference Models (SISO-STD-006-2010),* SISO COTS Simulation Package Interoperability Product Development Group.. Simulation Interoperability Standards Organization, Inc., P.O. Box 781238, Orlando, FL 32878-1238, USA. Information available on line via: http://www.sisostds.org/DigitalLibrary.aspx?Command=Core_Download&EntryId=30829 [last accessed: Jan. 19, 2011].

116. Smith, R., 2007. The Disruptive Potential of Game Technologies: Lessons Learned from its Impact on the Military Simulation Industry. *Research Technology Management.* Vol. 50, No. 2, 57-64.

117. Sriram, R.D., R. Sudarsan, X. Fiorentini, and S. Ray, 2009. Towards a Method for Harmonizing Information Standards. *5th Annual IEEE Conference on Automation Science and Engineering*, Paper MoC4.1, August 22-25, 2009, Bangalore, India.

118. Steinhauser, M.O., 2008. *Computational Multiscale Modeling of Fluids and Solids: Theory and Applications*, Springer-Verlag Berlin Heidelberg.

119. Tako, A. A., and S. Robinson, 2009. Comparing discrete-event simulation and system dynamics : users' perceptions. *The Journal of the Operational Research Society*, Vol. 60, No. 3, 296-312.

120. Taylor, J.G. and B. Neta, 2001. Support of JCATS Limited V&V. Naval Postgraduate School, Report no. NPS-MA-01-001. Available on line via: http://www.math.nps.navy.mil/~bneta/JCATSVV.pdf [last accessed Jan. 12, 2009].

121. Taylor, S.J.E., N. Mustafee, S. Kite, S. Strassburger, S.J. Turner, and C. Wood, 2010. Improving Modeling and Simulation Through Advanced Computing Techniques: Grid Computing and Distributed Simulation. *Proceedings of the 2010 Winter Simulation Conference*, Eds: B. Johansson, S. Jain, J. Montoya-Torres, J. Hugan, and E. Yücesan. 216-230. Piscataway, NJ: Institute of Electrical and Electronics Engineers.

122. Teilans, A., A. Kleins, Y. Merkuryev, and A. Grinbergs, 2008. Design of UML models and their simulation using ARENA. *WSEAS Transactions on Computer Research*, Vol. 3, Issue 1, January, 67-73.

123. Thacker, B.H., S.W. Doebling, F.M. Hemez, M.C. Anderson, J.E. Pepin, and E.A. Rodriguez, 2004. Concepts of Model Verification and Validation. Report LA-14167-MS. Los Alamos National Laboratory, Los Alamos, NM.

124. Tichenor, S. and A. Reuther, 2006. Making the Business Case for High Performance Computing: A Benefit-Cost Analysis Methodology. *CTWatch Quarterly*, Vol. 2 No. 4A, November 2006 A. Available on-line via: http://www.ctwatch.org/quarterly/articles/2006/11/making-the-business-case-for-high-performance-computing-a-benefit-cost-analysis-methodology/1/index.html [last accessed Sept. 10, 2009].

125. Trucano, T.G. and J.L. Moya, 1999. Guidelines for Sandia ASCI Verification and Validation Plans - Content and Format: version 1.0. Report SAND99-3098. Sandia National Laboratories, Albuquerque, NM.

126. Turnitsa, C., 2005. Extending the Levels of Conceptual Interoperability Model. *Proceedings of 2005 Summer Simulation Conference*. Cherry Hill, N.J., July 24-28. Society for Computer Simulation. ISBN-10: 1565552997.

127. USGSA, 2009. Section 508. U.S. General Services Administration. Available on-line at: http://www.section508.gov/index.cfm?FuseAction=Content&ID=9 [last accessed on April 18, 2009].

128. Vaughn, A., 2008. Simulation and Analysis of Border Operations. Presentation to DHS/NIST team, April 22, 2008.

129. Visarraga, D., 2008. Interdependency Environment for Infrastructure System Simulations (IEISS). National Infrastructure Simulation & Analysis Center (NISAC). Presentation No. LA-UR-07-3878. Provided to NIST team on Aug. 27, 2008.

130. White, W.J., A.C. O'Connor, and B.R. Rowe, 2004. Economic Impact of Inadequate Infrastructure for Supply Chain Integration. National Institute of Standards and Technology (NIST) Planning Report 04-2.

131. WMO, 2007. *Binary Universal Form for Data Representation of Meteorological Data (BUFR)*. World Meteorological Organization (WMO). Available on-line via: http://www.wmo.int/pages/prog/www/WMOCodes/Operational/BUFR/FM94REG-11-2007.pdf

132. NWS, 2009. Digital Weather Markup Language (DWML). National Weather Service (NWS). Available on-line via: www.nws.noaa.gov/mdl/XML/Design/MDL_XML_Design.ppt [last accessed on October 8, 2009].

133. Xie, J., C. Yang, B. Zhou and Q. Huang, 2009. High-performance computing for the simulation of dust storms. *Computers, Environment and Urban Systems*. Article in Press. doi:10.1016/j.compenvurbsys.2009.08.002

134. Zeigler, B.P., 1976. *Theory of Modelling and Simulation*. Wiley, New York.

Appendix A: Organizations visited for this study

Listed in alphabetical order.

1. Argonne National Labs, Chicago, IL
2. Applied Physics Lab, Johns Hopkins University, Laurel, MD
3. Breakaway Games, Hunt Valley, MD
4. Department of Homeland Security, Washington, DC
5. District Department of Transportation, Washington, DC
6. Lawrence Berkeley National Labs, Berkeley, CA
7. Lawrence Livermore National Labs, Livermore, CA
8. Los Alamos National Labs, Los Alamos, NM
9. Naval Postgraduate School, Monterey, CA
10. Sandia National Labs, Livermore, CA
11. Sandia National Labs, Albuquerque, NM
12. Transportation Security Administration, DHS, Arlington, VA
13. US Coast Guard, Washington, DC
14. US Fire Administration, Emmitsburg, MD
15. US Secret Service Training Facility, Laurel, MD

Appendix B: List of Acronyms

ANL	Argonne National Laboratory
CIPDSS	Critical Infrastructure Protection Decision Support System
DHS	Department of Homeland Security
EPA	Environmental Protection Agency
IMAAC	Interagency Modeling and Atmospheric Assessment Center
LANL	Los Alamos National Laboratory
LBNL	Lawrence Berkeley National Laboratory
LLNL	Lawrence Livermore National Laboratory
M&S	Modeling and Simulation
MS&A	Modeling, Simulation and Analysis
NARAC	National Atmospheric Release Advisory Center
NISAC	National Infrastructure Simulation and Analysis Center
NOAA	National Oceanic and Atmospheric Administration
NRC	Nuclear Regulatory Commission
PACER	National Center for the Study of Preparedness and Catastrophic Event Response, a DHS University Center of Excellence (UCE) led by Johns Hopkins University.
Sandia	Sandia National Laboratory
V&V	Verification and Validation
VV&A	Verification, Validation and Accreditation